Optical Fiber Applications

Edited by Guillermo Huerta-Cuellar
and Roghayeh Imani

Published in London, United Kingdom

IntechOpen

Supporting open minds since 2005

Optical Fiber Applications
http://dx.doi.org/10.5772/intechopen.83272
Edited by Guillermo Huerta-Cuellar and Roghayeh Imani

Contributors
Ulrich H.P. Fischer-Hirchert, Mladen Joncic, Richard Phelan, Diarmuid Byrne, Michael Gleeson, John O'Carroll, Marta Nawrocka, Rob Lennox, Kevin Carney, Chris Herbert, Jim Somers, Brian Kelly, Hamdy H. Wahba, Wael Ramadan, Ahmed El-Tawargy, Muhammad Irfan, Igor Chychura, Rafael Sanchez-Lara, Lelio De La Cruz-May, Jose Luis Vazquez-Avila, Daniel E. Ceballos-Herrera, Jose A. Alvarez-Chavez, Guillermo Huerta-Cuellar, Roghayeh Imani

First published in London, United Kingdom, 2020 by IntechOpen
IntechOpen is the global imprint of INTECHOPEN LIMITED, registered in England and Wales, registration number: 11086078, 7th floor, 10 Lower Thames Street, London, EC3R 6AF, United Kingdom
Printed in Croatia

British Library Cataloguing-in-Publication Data
A catalogue record for this book is available from the British Library

Additional hard and PDF copies can be obtained from orders@intechopen.com

Optical Fiber Applications
Edited by Guillermo Huerta-Cuellar and Roghayeh Imani
p. cm.
Print ISBN 978-1-78985-615-6
Online ISBN 978-1-78985-616-3
eBook (PDF) ISBN 978-1-83880-374-2

For EU product safety concerns:
IN TECH d.o.o., Prolaz Marije Krucifikse Kozulić 3, 51000 Rijeka, Croatia, info@intechopen.com or visit our website at intechopen.com.

Meet the editors

Dr. Guillermo Huerta-Cuellar completed his doctoral studies at the Research Center in Optics, and a postdoctoral stay at the Potosino Institute of Scientific and Technological Research. He is currently Tenure Professor-Researcher at the University of Guadalajara, Mexico, is recognized for the desirable profile PRODEP, and is a member of both the National System of Researchers Level I and the consolidated research group Complex Systems, Optics and Innovation. Dr. Huerta-Cuellar has more than eight years' teaching experience at undergraduate, masters, and doctorate levels. He has made research stays in Mexico, Italy, Spain, Russia, and the United States of America. He has produced 31 indexed articles, five book chapters, and 12 congress articles. To date, he has directed and completed six theses at undergraduate level and two at masters level, and has a PhD in progress.

Dr. Roghayeh Imani is a semiconductor physicist. Her main expertise is in nanostructured semiconductor growth, characterization, and optoelectronic devices. Parallel to experimental research, she is also studying computer modeling and simulation to design and predict new types of semiconductors. She received her bachelor and master degrees in physics, and obtained her undergraduate training at the Laboratory of Thin Film and Surface Physics. She was awarded her doctoral degree in Nanoscience-Physics in November 2015 from the Electrical Engineering Faculty of Ljubljana University. During her PhD program she conducted impressive research into the biomedical application of multifunctional nanostructured semiconductors, and in 2017 her PhD was selected as a top-10 research subject in Ljubljana University. So far, her postdoctoral research in Sweden has been focused on the development of optoelectronic devices based on nanostructured semiconductors.

Contents

Preface

After the invention of the laser in the 1960s, the appearance of optical fibers enhanced applications that required the transportation of light from one place to another. As is well known, one of the main applications of optical fibers is optical communications, which were initially inefficient. Over time, different groups of scientists and engineers have been working to improve the efficiency of information transmission in optical fibers with materials that reduce absorption to the greatest extent possible. However, a large number of applications have been developed for which light absorption is not a real problem since the distances for sending information can be very short. Therefore, the technological development of optical fibers, for short or long distances, has been applied to a variety of uses in different areas such as the armed forces, industry, medicine, scientific research, and of course in communications, to name a few.

Among the large number of applications that exist in optical fibers are different types of sensors, medical equipment to perform invasive surgeries, devices for industrial applications used in difficult access situations, various toys, decorative lighting, long-distance communications, and optical amplifiers, among others. Because of their great potential in applications and the advantages they offer over other systems, optical fibers will continue to be the subject of scientific research and technological development.

Guillermo Huerta Cuellar
Dynamical Systems Laboratory,
Universidad de Guadalajara,
Centro Universitario de los Lagos,
Lagos de Moreno,
Jalisco, México

Applied Mathematics Division,
Instituto Potosino de Investigación Científica y Tecnológica, IPICYT,
San Luis Potosí, SLP,
México

Roghayeh Imani
Structural Chemistry,
Department of Chemistry-Ångström Laboratory,
Uppsala University,
Uppsala, Sweden

Section 1

Introduction

Chapter 1

Introductory Chapter: Optical Fibers

Roghayeh Imani and Guillermo Huerta Cuellar

1. Optical fibers

The optical fibers invented by Kapany [1], based on the observations that John Tyndall made a few years before, have become the object of research, technological development, and applications to this day.

The optical fibers have the ability to transport a large amount of information between two points (emitter-receiver). Unlike the conductive cables that are commonly used for sending information, fiber-optic cables offer the advantage of being very light [2]. As for its physical characteristics, an optical fiber is commonly compared to a human hair whose diameter is around 120 μm [3]. In terms of capacity for sending information, they can carry up to 20 billion light pulses per second [3]. Nowadays, due to the high transmission capacity and low absorption losses in an optical fiber, it is possible to send information over distances of more than 100 km without the need for repeaters. To carry out the sending of information from one point to another by means of a fiber-optic system, three basic elements should be considered as mentioned in **Figure 1**:

- A transmitter, which generates the wave signal to be transmitted, which is fed with the information to be sent
- Fiber-optic cable, which corresponds to the medium in which the information moves from the sender to the receiver
- Receiving system, which receives the information sent

In order to have mechanical support and total reflection of the light traveling in the fiber [4], avoiding absorption, the optical fiber is composed of three main elements (**Figure 2**):

- The core of the fiber, which is the fiber-optic glass where the light to be transmitted moves
- The cladding, which consists of a thin layer of material whose purpose is to achieve light reflection in the fiber and prevent it from escaping, based on the principle of total internal reflection
- The coating, made of plastic, similar to that used on copper cables, which provides protection to the fiber against climatic factors, corrosion, or external damage

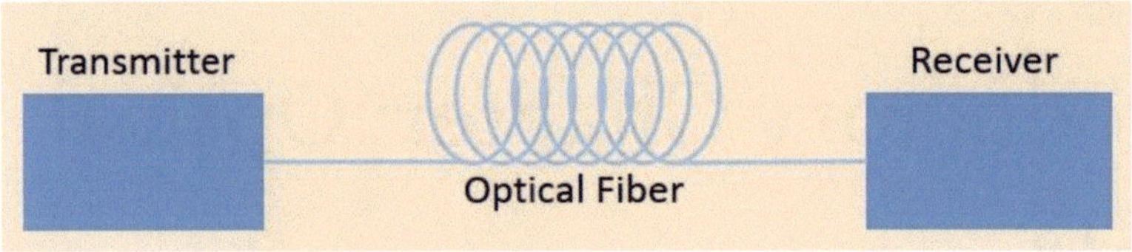

Figure 1.
Basic schema of an optical communication system.

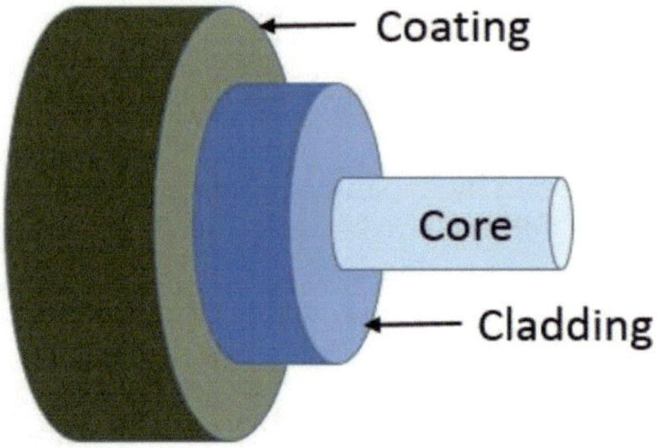

Figure 2.
Principal elements of an optical fiber.

Due to their ease of manufacturing, the optical fibers are made of glass and plastic; however due to their performance, the optical glass fibers are the most used due to the transmission length and their efficiencies. Basically there are two types of optical fibers, and they are defined by their ability to transmit information, which implies the type of application [4]. Those fiber types are mentioned next:

1. Single-mode fibers: they have a smaller diameter than multimode fibers. In this type of fibers, the light travels parallel to the axis which creates a small dispersion. In these fibers the modes of transmission are many, and therefore the distances that these fibers can cover can be more than 50 times those of multimodal fibers. These types of fibers are the most used by communications companies.

2. Multimode fibers: they can send different data transmissions simultaneously on a single fiber. Its diameter is slightly larger than that of single-mode fibers, and it can be 50 and 62.5 μm, which allows light to enter at different angles.

Figure 3 shows a graphic comparison of the expected behavior in the two types of the mentioned fibers; the way of sending information between the fibers can be noted.

When the copper transmission capacity comparison is made, which allows a few million pulses per second, versus more than 20 million pulses that can be sent in an optical fiber, it is possible to appreciate the great efficiency advantage. It is understandable that communications companies can work with large amounts of information that would be limited with the use of conductive cables. Therefore, the great increase in the transmission of information that we can see on the Internet and the Word Wide Web has been achieved [5].

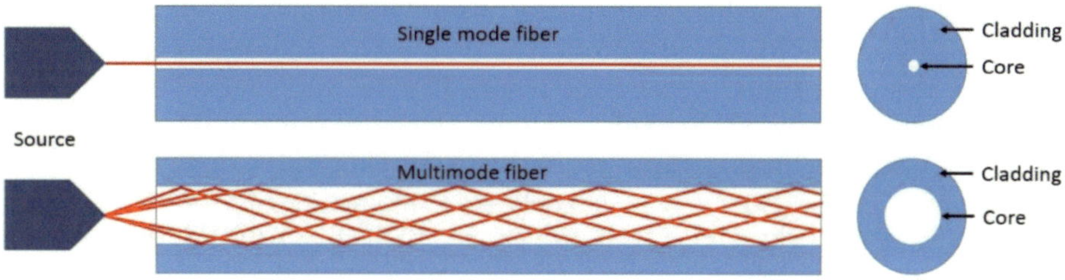

Figure 3.
Simple graphical comparison between single-mode and multimode optical fibers.

2. Optical fiber transmission

From the invention of optical fibers to the present day, research and technological development on optical fibers has been carried out [6]. The use of semiconductor devices coupled to fiber-optics in technological development for the transmission of information has evolved over time since 1980. Five generations of technological development have been implemented where fiber-optic technology coupled to semiconductor materials has increased the amount of information as well as the distance between repeaters (**Table 1**).

Generation	Type	Year	Bit rate	Repeater spacing	Operating wavelength
First	Graded-index fibers	1980	45 Mb/s	10 km	0.8 μm
Second	Single-mode fibers	1985	100 Mb/s to 1.7 Gb/s	50 km	1.3 μm
Third	Single-mode lasers	1990	10 Gb/s	100 km	1.55 μm
Fourth	Optical amplifiers	1996	10 Tb/s	>10,000	1.45–1.62 μm
Fifth	Raman amplification	2002	40–160 Gb/s	24,000–35,000 km	1.53–1.57 μm

Table 1.
Evolution of different generations of light wave systems.

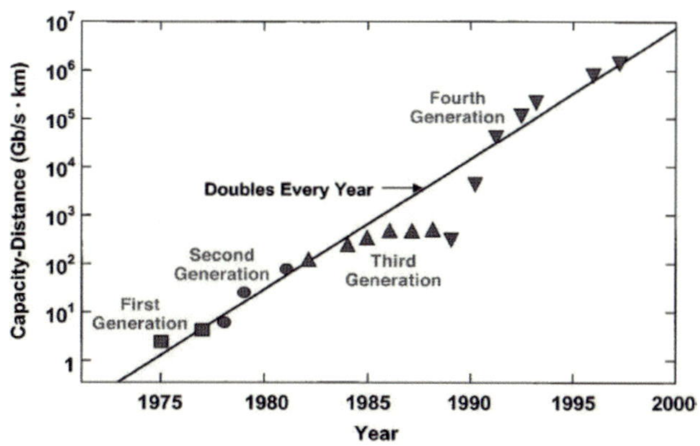

Figure 4.
Increase in the BL product from 1975 to 2000 through four generations of light wave systems. Different symbols are used for successive generations [4].

Figure 4 shows the evolution of the first four generations of optical fiber coupled with laser devices as an emission source. Two mechanisms can be mentioned about the fourth generation to amplify the information that reaches a system; these are the rare earth-doped amplifiers and semiconductor amplifiers [7, 8]. Additionally as mentioned in the fifth generation, there are Raman amplifiers [9].

Among the amplifiers doped with rare earths, those made with fiber-optic doped with erbium and doped with ytterbium can be mentioned [10–12]. The optical fibers doped with erbium are the most used for the construction of this type of amplifiers, among other sensors and other applications [13–16]. The above due to its low cost compared to the ytterbium and because its amplification window coincides with the third transmission window of silica-based optical fibers [17].

3. Optical fiber applications

Regarding the advantages of optical fibers for the transmission of information and their multiple applications over existing methods such as those based on conductors such as copper, the following can be mentioned:

1. Perhaps the most outstanding feature of optical fibers is that they are capable of transmitting at the speed of light, which greatly increases the ability to send information.

2. Optical fibers are not affected by electrical signals that distort information, as is the case with conductive cables.

3. Due to the absence of intermittency due to electromagnetic fields, a fiber-optic cable cannot be detected remotely since it can only be accessed by entering the fiber. This is very attractive as a security measure for the government, banks, and other companies that need to protect their information.

4. Unlike metallic conductors, optical fibers are not capable of producing sparks with the information they carry; therefore, there is no risk of fire due to such events.

5. The signal absorption in the optical fiber is much lower than in a copper cable, in addition to the fact that cladding allows a very efficient internal reflection. With this, in applications where information is transmitted over several channels, there is no possibility of mixing data between fibers as can occur in cable communications.

6. The cost of installing several kilometers of fiber-optic is cheaper than the installation of copper cables. In this way, communication services such as the Internet are cheaper because with low-power intensity, information can be transmitted over long distances.

7. The bandwidth that can be had, as well as the data capacity of an optical fiber, allows to have a lot of information in a very thin cable. This is the reason for the continuous research and technological innovation for its increase.

8. Fiber-optic cable installations are less expensive than copper or coaxial cable installations, as well as the installation equipment.

9. It is well known that the distances for transmission through the use of optical fiber are very long, more than 100 km without active or passive elements intervening, in addition to the obtained low attenuation.

Additionally, some disadvantages could have been found; perhaps one of the main ones is that optoelectronic devices that can be connected are expensive.

Comparing the emitting and receiving devices of electrical systems against those used with fiber-optic technology, this difference can be understood. That is why some companies still use electrical devices.

Considering the advantages, simplicity of manufacturing, as well as its versatility and growth in popularity, multiple technological applications and practical uses for optical fibers have been developed.

Some of the optical fiber applications which are not sensitive to electrical or magnetic interference make them highly recommended for military applications [18–20]. In addition there are different applications in industrial areas, such as sensing an ambient temperature, pressure, and gases or even wiring of automotive systems [21–24].

In the area of medicine, optical fibers are the modifications for laser surgery, such as light guides and tools for the transmission of images [25–28]. In the area of sciences, fiber-optic applications are very useful [21, 25]. In the manipulation of particles by means of optical tweezers, the optical fibers provide a great advantage [29].

In the subject of communications, the use of optical amplifiers has been outstanding over other applications. These optical amplifiers have been developed by manufacturing doped optical fibers with rare earth ions to have light amplification by stimulated emission [4, 30, 31]. **Table 2** shows some of the most common dopants, as well as the glass used as the host for their application.

In addition, systems with nonlinear behavior have been found. Among the results observed with the said nonlinear behavior in fiber doped with erbium (Er^{3+}), there are behaviors with multistability, phenomenological dynamics, different sensors, and masking in communications, among others [14, 16, 23, 33–35].

As mentioned above, to improve all the possible applications in a fiber-optic system, the process of integration with optoelectronic devices is necessary.

This book offers a comprehensive review about the design, manufacturing, and performance obtained from the integration of fiber-optics with optoelectronic devices. In particular, it analyzes some of the advances in light-emitting diodes (LEDs), lasers, photodetectors, as well as applications in communication systems, sensors, interferometric, and holographic methods, which with the use of optical fiber strengthens the capacity and applications.

Dopant	Common host glasses	Important emission wavelengths
Erbium (Er^{3+})	Silicate and phosphate glasses, fluoride glasses	1.51.6 μm, 2.7 μm, 0.55 μm
Holmium (Ho^{3+})	Silicate glasses, fluorozirconate glasses	2.1 μm, 2.9 μm
Neodymium (Nd^{3+})	Silicate and phosphate glasses	1.031.1 μm, 0.90.95 μm, 1.321.35 μm
Praseodymium (Pr^{3+})	Silicate and fluoride glasses	0.3 μm, 0.635 μm, 0.6 μm, 0.52 μm, 0.49 μm
Thulium (Tm^{3+})	Silicate and germanate glasses, fluoride glasses	1.72.1 μm, 1.451.53 μm, 0.48 μm, 0.8 μm
Ytterbium (Yb^{3+})	Silicate glass	1.01.1 μm

Table 2.
Common laser-active ions, host glasses, and important emission wavelength [32].

Acknowledgements

G.H.C. acknowledges E. Campos-Cantón for the opportunity to realize a research stay with his investigation group at IPICYT and the financial support from the University of Guadalajara under the projects Research Laboratory Equipment for Academic Groups in Optoelectronics from CULAGOS, R-0138/2016; Agreement RG/019/2016-UdeG and RC/075/2018; and Agreement RG/006/2018, UDG, México.

Author details

Roghayeh Imani[1] and Guillermo Huerta Cuellar[2,3]*

1 Department of Engineering Sciences and Mathematics, Luleå University of Technology, Luleå, Sweden

2 Dynamical Systems Laboratory, Centro Universitario de los Lagos, Universidad de Guadalajara, Lagos de Moreno, Jalisco, Mexico

3 Applied Mathematics Division, Instituto Potosino de Investigación Científica y Tecnológica (IPICYT), San Luis Potosí, Mexico

*Address all correspondence to: g.huerta@lagos.udg.mx

References

[1] Kapany NS. Fiber optics. Scientific American. 1960;**203**(5):72-81

[2] Paradisi A, Carvalho Figueiredo R, Chiuchiarelli A, de Souza Rosa E. Optical Communications. Switzerland: Springer Press; 2019. Available from: https://www.springer.com/la/book/9783319971865

[3] Hoss R, Ptaschek L. Manufacturing methods and technology program for ruggedized tactical fiber optic cable. ITT Electro-optical Products div roanoke va. 1979. Available from: https://apps.dtic.mil/docs/citations/ADA084191

[4] Agrawal GP. Fiber-Optic Communication Systems. Vol. 222. New Jersey, Canada: John Wiley & Sons; 2012. Available from: https://www.wiley.com/enus/Fiber+Optic+Communication+Systems,+4th+Edition-p-9780470505113

[5] Lopez O, Haboucha A, Chanteau B, Chardonnet C, Amy-Klein A, Santarelli G. Ultra-stable long distance optical frequency distribution using the internet fiber network. Optics Express. 2012;**20**(21):23518-23526. Available from: https://www.osapublishing.org/oe/abstract.cfm?uri=oe-20-21-23518

[6] Shumate PW, Snelling RK. Evolution of fiber in the residential loop plant. IEEE Communications Magazine. 1991;**29**(3):68-74. Available from: https://ieeexplore.ieee.org/abstract/document/75529

[7] Durhuus T, Mikkelsen B, Joergensen C, Danielsen SL, Stubkjaer KE. All-optical wavelength conversion by semiconductor optical amplifiers. Journal of Lightwave Technology. 1996;**14**(6):942-954. Available from: https://ieeexplore.ieee.org/abstract/

[8] Tanabe S. Optical transitions of rare earth ions for amplifiers: How the local structure works in glass. Journal of Non-Crystalline Solids. 1999;**259**(1-3): 1-9. Available from: https://www.sciencedirect.com/science/article/abs/pii/S0022309399004901

[9] Islam MN. Raman amplifiers for telecommunications. IEEE Journal of Selected Topics in Quantum Electronics. 2002;**8**(3):548-559. Available from: https://ieeexplore.ieee.org/abstract/document/1016358

[10] Barbier D, Rattay M, Saint Andre F, Clauss G, Trouillon M, Kevorkian A, et al. Amplifying four-wavelength combiner, based on erbium/ytterbium-doped waveguide amplifiers and integrated splitters. IEEE Photonics Technology Letters. 1997;**9**(3):315-317. Available from: https://ieeexplore.ieee.org/abstract/document/556058

[11] Kir'Yanov AV, Barmenkov YO, Sandoval-Romero GE, Escalante-Zarate L. Er^{3+} concentration effects in commercial erbium-doped silica fibers fabricated through the MCVD and DND technologies. IEEE Journal of Quantum Electronics. 2013;**49**(6):511-521. Available from: https://ieeexplore.ieee.org/abstract/document/6502181

[12] Kir'yanov AV, Paul MC, Barmenkov YO, Das S, Pal M, Escalante-Zarate L. Yb^{3+} concentration effects in novel Yb doped lanthano-alumino-silicate fibers: Experimental study. IEEE Journal of Quantum Electronics. 2013;**49**(6):528-544. Available from: https://ieeexplore.ieee.org/abstract/document/6506094

[13] Escalante-Zarate L, Barmenkov YO, Kolpakov SA, Cruz JL, Andrés MV. Smart Q-switching for single-pulse generation in an erbium-doped fiber laser. Optics Express.

2012;**20**(4):4397-4402. Available from: https://www.osapublishing.org/oe/abstract.cfm?uri=oe-20-4-4397

[14] García-López JH, Jaimes-Reátegui R, Afanador-Delgado SM, Sevilla-Escoboza R, Huerta-Cuellar G, López-Mancilla D, et al. Experimental and numerical study of an optoelectronics flexible logic gate using a chaotic doped fiber laser. Recent Development in Optoelectronic Devices. 2018;**97**:97-114. Available from: https://www.intechopen.com/books/recent-development-in-optoelectronic-devices

[15] Jaimes-Reátegui R, Afanador-DelgadoSM,Sevilla-EscobozaR, Huerta-Cuellar G, García-López JH, López-Mancilla D, et al. Optoelectronic flexible logic gate based on a fiber laser. The European Physical Journal Special Topics. 2014;**223**(13):2837-2846. Available from: https://link.springer.com/article/10.1140/epjst/e2014-02297-4

[16] Pisarchik AN, Sevilla-Escoboza R, Jaimes-Retegui R, Huerta-Cuellar G, Garca-Lopez JH, Kazantsev VB. Experimental implementation of a biometric laser synaptic sensor. Sensors. 2013;**13**(12):17322-17331. Available from: https://www.mdpi.com/1424-8220/13/12/17322

[17] Chapman DA. Erbium-doped fibre amplifiers: The latest revolution in optical communications. Electronics & Communication Engineering Journal. 1994;**6**(2):59-67. Available from: https://digital-library.theiet.org/content/journals/10.1049/ecej19940202

[18] García I, Zubia J, Durana G, Aldabaldetreku G, Illarramendi MA, Villatoro J. Optical fiber sensors for aircraft structural health monitoring. Sensors. 2015;**15**(7):15494-15519. Available from: https://www.mdpi.com/1424-8220/15/7/15494

[19] Saito N, Yari T, Hotate K, Kishi M, Matsuura S, Kumagai Y, et al. Developmental status of SHM applications for aircraft structures using distributed optical fiber. Structural Health Monitoring. 2013;**2**:2011-2018. Available from: http://www.dpi-proceedings.com/index.php/shm2013/article/view/23012

[20] Makovejs S, Roberts CC, Palacios F, Matthews HB, Lewis DA, Smith DT, et al. Record-low (0.1460 dB/km) attenuation ultra-large Aeff optical fiber for submarine applications. In: 2015 Optical Fiber Communications Conference and Exhibition (OFC). 2015. pp. 1-3. Available from: https://ieeexplore.ieee.org/abstract/document/7122138

[21] Lee B. Review of the present status of optical fiber sensors. Optical Fiber Technology. 2003;**9**(2):57-79. Available from: https://www.sciencedirect.com/science/article/pii/S1068520002005278

[22] Monzon-Hernandez D, Martinez-Rios A, Torres-Gomez I, Salceda-Delgado G. Compact optical fiber curvature sensor based on concatenating two tapers. Optics Letters. 2011;**36**(22):4380-4382. Available from: https://www.osapublishing.org/ol/abstract.cfm?uri=ol-36-22-4380

[23] Barmenkov YO, Mendoza-Santoyo F. Faraday plasma current sensor with compensation for reciprocal birefringence induced by mechanical perturbations. Journal of Applied Research and Technology. 2003;**1**(2):157-163. Available from: http://www.scielo.org.mx/scielo.php?pid=S1665-64232003000200006&script=sci_arttext&tlng=en

[24] Willer U, Saraji M, Khorsandi A, Geiser P, Schade W. Near-and mid-infrared laser monitoring of industrial processes, environment and security applications. Optics and Lasers in Engineering. 2006;**44**(7):699-710. Available from: https://www.

sciencedirect.com/science/article/abs/pii/S0143816605001107

[25] Mishra V, Singh N, Tiwari U, Kapur P. Fiber grating sensors in medicine: Current and emerging applications. Sensors and Actuators A: Physical. 2011;**167**(2):279-290. Available from: https://www.sciencedirect.com/science/article/abs/pii/S092442471100118X

[26] Taffoni F, Formica D, Saccomandi P, Pino GD, Schena E. Optical fiber-based MR-compatible sensors for medical applications: An overview. Sensors. 2013;**13**(10):14105-14120. Available from: https://www.mdpi.com/1424-8220/13/10/14105/htm

[27] Song H, Kim K, Lee J. Development of optical fiber Bragg grating force-reflection sensor system of medical application for safe minimally invasive robotic surgery. Review of Scientific Instruments. 2011;**82**(7):074301. Available from: https://aip.scitation.org/doi/full/10.1063/1.3606502

[28] Russo V, Righini GC, Sottini S, Trigari S. Lens-ended fibers for medical applications: A new fabrication technique. Applied Optics. 1984;**3**(19):3277-3283. Available from: https://www.osapublishing.org/ao/abstract.cfm?uri=ao-23-19-3277

[29] Gelfand RM, Wheaton S, Gordon R. Cleaved fiber optic double nanohole optical tweezers for trapping nanoparticles. Optics Letters. 2014;**39**(22):6415-6417. Available from: https://www.osapublishing.org/ol/abstract.cfm?uri=ol-39-22-6415

[30] Tanabe S. Rare-earth-doped glasses for fiber amplifiers in broadband telecommunication. Comptes Rendus Chimie. 2002;**5**(12):815-824. Available from: https://www.sciencedirect.com/science/article/abs/pii/S1631074802014492

[31] Huerta-Cuellar G, Pisarchik AN, Kiryanov AV, Barmenkov YO, del Valle Hernndez J. Prebifurcation noise amplification in a fiber laser. Physical Review E. 2009;**79**(3):036204. Available from: https://journals.aps.org/pre/abstract/10.1103/PhysRevE.79.036204

[32] Poole S, Payne D, Mears R, Fermann M, Laming R. Fabrication and characterization of low-loss optical fibers containing rare-earth ions. Journal of Lightwave Technology. 1986;**4**(7):870-876. Available from: https://ieeexplore.ieee.org/abstract/document/1074811

[33] Huerta-Cuellar G, Pisarchik AN, Barmenkov YO. Experimental characterization of hopping dynamics in a multistable fiber laser. Physical Review E. 2008;**78**(3):035202. Available from: https://journals.aps.org/pre/abstract/10.1103/PhysRevE.78.035202

[34] Pisarchik AN, Jaimes-Retegui R, Sevilla-Escoboza R, Huerta-Cuellar G, Taki M. Rogue waves in a multistable system. Physical Review Letters. 2011;**107**(27):274101. Available from: https://journals.aps.org/prl/abstract/10.1103/PhysRevLett.107.274101

[35] Pisarchik AN, Jaimes-Retegui R, Sevilla-Escoboza R, Huerta-Cuellar G. Multistate intermittency and extreme pulses in a fiber laser. Physical Review E. 2012;**86**(5):056219. Available from: https://journals.aps.org/pre/abstract/10.1103/PhysRevE.86.056219

Section 2

Recent Optoelectronic Devices and Light Transmission

Chapter 2

Mid-Infrared InP-Based Discrete Mode Laser Diodes

Richard Phelan, Diarmuid Byrne, John O'Carroll, Michael Gleeson, Marta Nawrocka, Rob Lennox, Kevin Carney, Chris Herbert, Jim Somers and Brian Kelly

Abstract

Low cost, compact and robust single mode semiconductor laser diodes emitting at $\lambda \sim 1.6$–2.1 μm are highly desirable as light sources for trace gas spectroscopy and an increasing number of other applications, such as, high data-rate communications over hollow core photonic crystal fibre, noninvasive optical blood glucose monitoring. Indium phosphide based light sources provide a solid and flexible base for mid-infrared semiconductor diode lasers. This chapter provides an overview of the current state of the art in discrete mode InGaAs/InP long-wavelength quantum-well lasers emitting in the 1.6–2.1 μm wavelength range. The discrete mode laser is essentially a regrowth free modified ridge waveguide Fabry-Pérot laser whose optical spectrum has a single wavelength mode. High-performance and cost-effective mid-infrared DM laser diode sources are well suited to a wide range of sensor applications. The current state of the art will also be outlined in this chapter.

Keywords: semiconductor laser, mid-infrared sources, single mode laser, strained quantum-well, absorption spectroscopy

1. Introduction

Low cost single mode semiconductor laser diodes emitting at wavelengths, λ, in the 1.6–2.1 μm wavelength range are highly desirable as light sources for trace gas spectroscopy due to the strong absorption bands [1] in this spectral region. Compared to conventionally used systems based on electrochemical point sensors, tuneable diode laser absorption spectroscopy (TDLAS) offers a number of benefits for the detection of gases such as; rapid response time, long term stability, high selectivity and sensitivity, rugged systems and measurement capability over long distances using open path systems [2]. TDLAS is based on the molecular rotational-vibrational absorption of gases that produce distinct peaks in the near to mid-infrared (IR) spectral range. These molecule resonances cause characteristic 'fingerprints' by selective absorption of laser light as the wavelength is tuned by changing the laser bias current or heat sink temperature. Great interest is due to the strong absorption lines of various important gases, such as methane (1.665 μm), hydrogen chloride (1.743 μm), and nitrous oxide (1.795 μm), in this wavelength range. Two pertinent greenhouse gases, water vapour (H_2O) and carbon dioxide (CO_2), have strong absorption bands at wavelengths centred around 1.877 and

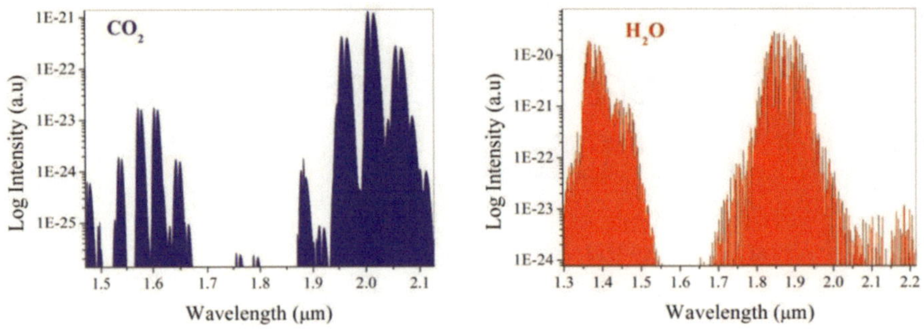

Figure 1.
Absorption spectra of two important greenhouse gases CO_2 and H_2O, with strong absorption lines in the near-to-mid-IR extracted from the HITRAN database [1].

2.004 μm respectively and are shown in **Figure 1**. There is also an increasing number of other applications, such as high data-rate communications over hollow core photonic crystal fibre [3–6] and noninvasive optical blood glucose monitoring [7] which also require compact and robust low-cost laser sources emitting in the 1.6–2.1 μm wavelength range.

Semiconductor laser diode materials with light emission in the 1.6–2.1 μm wavelength range include indium phosphide (InP) and the gallium antimonide (GaSb) material systems. It must be noted that cost sensitivity is a significant issue for many of these applications and in order to keep the laser chip cost down the InP material is often preferred over the GaSb material system for emission in the mid-IR region. In comparison with the GaSb material platform, the processing technologies for InP-based materials are more mature as they were developed for telecommunications lasers. In addition, superior substrate quality, lower substrate cost, better thermal performance and mature growth methods make InP-based lasers attractive candidates for light sources in this wavelength region [8–15]. The III-V compound semiconductor material system (AlGaIn)-(AsP) constitutes an ideal basis for the realization of diode lasers in this wavelength region [16]. InGaAs, either lattice matched to InP or deliberately strained, can be used for the active layer with a direct band gap between 1.1 and 2.3 μm. Recently InP-based type-II QWs have extended the wavelength up to 3 μm [17] which opens up another important wavelength region for the fabrication of lower cost lasers for sensing applications.

This chapter begins with an overview of mid-IR single mode laser diodes and then outlines the state of the art in InP based mid-IR discrete mode laser diodes.

2. Overview of the state of the art in mid-IR single mode laser diodes

In this section an overview of the current state of the art in compact monolithic single mode laser diodes emitting in the 2.0 μm spectral region will be outlined. Focusing on monolithic chips, external cavity laser devices are not included here. The review will start with an overview of the material systems available for the laser active region emission in the 2 μm spectral region. To date single mode semiconductor lasers have been demonstrated in both the GaSb and indium InP material systems. Both material systems have their pros and cons and an overview of the various technology implementations for achieving single wavelength mode operation in each material system is shown.

2.1 Gallium antimonide (GaSb) laser diodes

The III-V compound semiconductor material system (AlGaIn)(AsSb) constitutes an alternative basis for the realisation of diode lasers in the mid-IR. GaInAsSb is either latticed matched to GaSb substrates or strained and is used as the active layer with a direct bandgap transition covering the $\lambda \sim 1.8$–3.4 μm wavelength region. For the barrier and cladding layers, AlGaAsSb is well suited because of its larger bandgap energy and lower refractive index compared to GaInAsSb. Molecular-beam epitaxy (MBE) is the method of choice and most widely used in the wafer growth. The laser structures are grown on (100)-orientated n-doped GaSb substrates but due to the lower demand for GaSb substrates compared to InP the growth cost is higher and substrate quality lower. Conventional fabrication of distributed-feedback (DFB) lasers incorporating buried gratings for longitudinal mode selection is challenging in the GaSb material system due to the difficulty of epitaxial regrowth with the high Al concentrations in the cladding layers. Separately, processing options are quite limited for GaSb due to the cladding material being hygroscopic.

2.1.1 Laterally coupled distributed-feedback

A proposed method for DFB fabrication which was recently demonstrated by the Jet Propulsion Laboratory (JPL) [18], makes use of Bragg gratings etched alongside a ridge waveguide to form a laterally coupled distributed-feedback (LC-DFB) as shown in **Figure 2** This approach enables fabrication of single-longitudinal-mode laser following a single epitaxial growth process.

Characteristics of the LC-DFB laser are as follows; at a heat sink temperature of −10°C they demonstrate a single mode emission in the 2054 nm region with an SMSR > 30 dB and ex-facet power exceeding 80 mW. This is the highest reported power from a DFB laser at 2 μm. The wavelength shift with current and temperature is reported to be 5.39×10^{-3} nm/mA and 0.2 nm/°C [19]. They also demonstrated LC-DFB laser linewidths of 1.4 MHz and 900 kHz for 500 and 10 ms observation times, respectively. The linewidths were derived from the frequency-noise power-spectral density measured using a Fabry-Perot interferometer.

2.1.2 Lateral metal grating DFB

Ridge waveguide GaSb DFB lasers have also been fabricated by the University of Wurzburg and commercialised by the German company NanoPlus which employ a lateral metal grating at the side of the ridge [20] as depicted in **Figure 3**. The metal gratings provide strong feedback but generate additional absorption loss in the laser cavity hence the ex-facet laser power is limited to ~10 mW levels.

A single mode emission at 2 μm with a side mode suppression ratio of 31 dB is demonstrated in [20] and output power in the 8 mW region. The wavelength tuning rate with temperature is 0.2 nm/°C. Reported linewidth measurements are in the 0.3–0.5 MHz region [21].

2.2 Indium phosphide (InP) laser diodes

To extend the emission wavelength from $\lambda \sim 1.55$–2.1 μm in the $In_xGa_{1-x}As$ material system compressive strain is applied by increasing the indium (In) composition. **Figure 6** shows the calculated bandgap wavelength (λ_g (μm) = $1.2407/E_g$ (eV)) for a bulk $In_xGa_{1-x}As$ layer on InP as a function of In composition. As shown

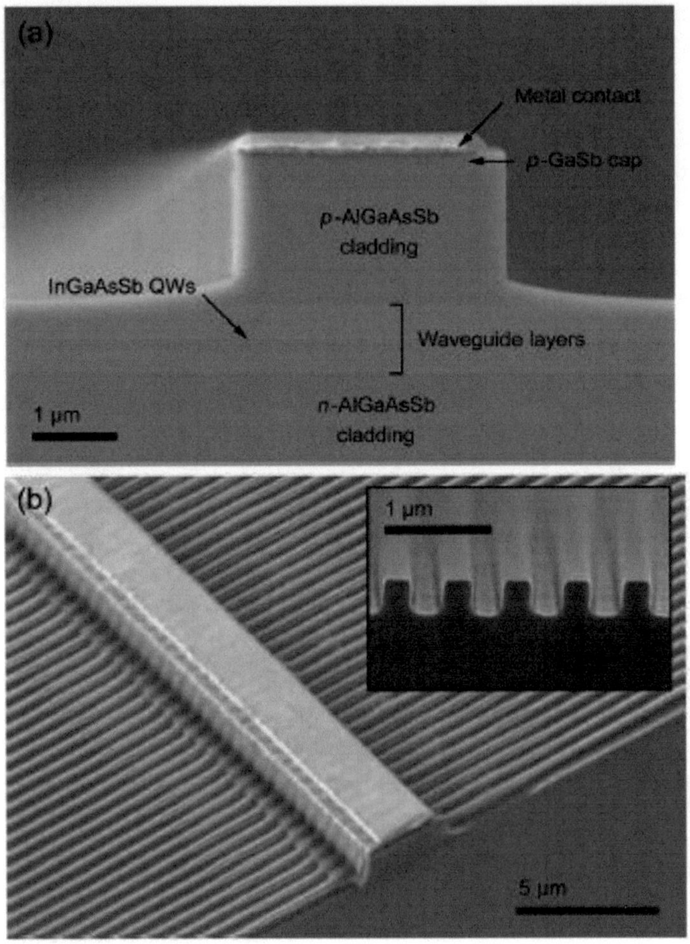

Figure 2.
(a) Cross-section scanning-electron micrograph of a 3-μm-wide laser ridge topped with a titanium-platinum-gold contact layer. (b) Top view of the LC-DFB laser structure, with a cross section of the grating (inset) [18].

in **Figure 2**, a compressive strain larger than 1% is required to obtain a bandgap wavelength longer than 2 μm. Also shown in **Figure 4** is the experimentally measured quantum well photoluminescence peak wavelength for four separate active regions with the In composition varied. When a quantum well structure is used instead of a bulk layer, the bandgap wavelength becomes smaller because of the quantum size effect on the bandgap energy. Therefore larger strain is required for $In_xGa_{1-x}As$ quantum wells (QWs) to obtain the same bandgap wavelength as bulk $In_xGa_{1-x}As$ [14].

2.2.1 InP-DFB

InP-based DFB lasers have been extensively developed with wavelengths at 1.3 and 1.55 μm for fibre optic communications over the last three decades. Similar processing techniques can be used for fabrication of single mode lasers operating in the 2 μm wavelength range. NTT-Japan demonstrated a DFB laser with an emission wavelength of 2.051 μm and output power of 10 mW [22]. A schematic of their buried heterostructure DFB grating is shown in **Figure 5**. The DFB grating was buried and required two regrowth stages after grating formation. There are no reported measurements on the spectral linewidths but they are expected to be in the

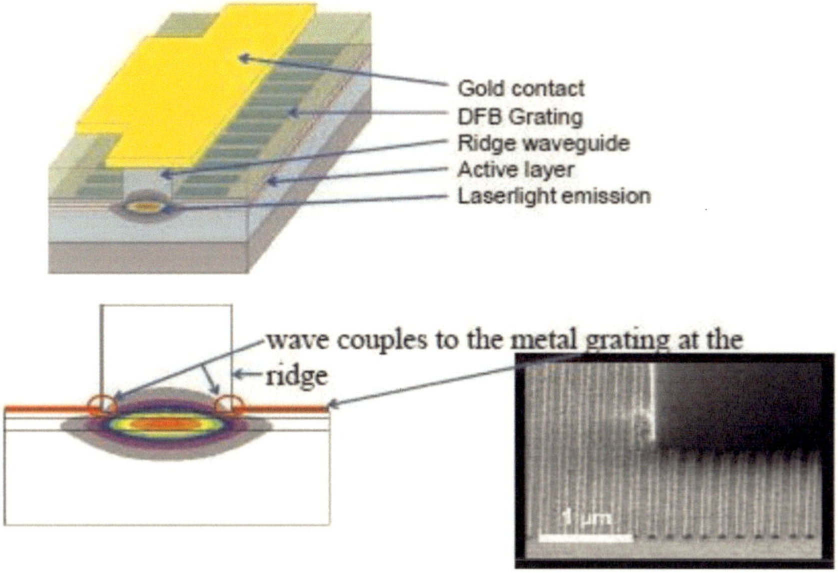

Figure 3.
Schematic diagram of the ridge waveguide metal grating DFB [20].

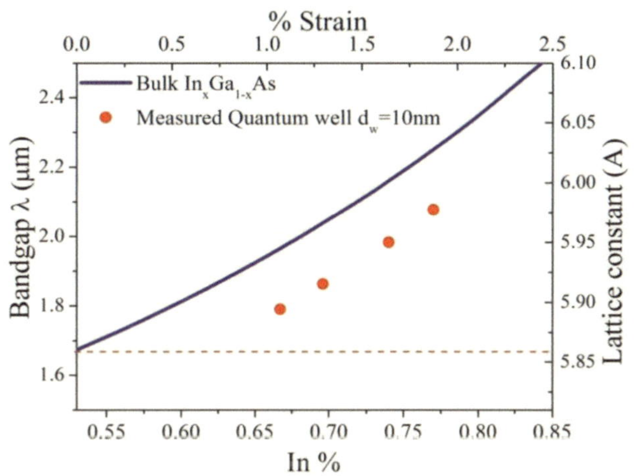

Figure 4.
Calculated bandgap wavelength for InGaAs on InP as a function of In composition. The upper horizontal axis shows the mismatch strain of InGaAs with respect to InP. Red dashed line indicates the lattice constant of InP 5.869A [13].

2 MHz range. The ex-facet power of 10 mW was improved on in a subsequent paper [23] with a value of ~20 mW reported. The wavelength shift with current and temperature is reported to be 0.0025 nm/mA and 0.125 nm/°C [24].

2.2.2 Vertical-cavity surface-emitting laser

Light propagation and emission normal to the semiconductor layer structure is characteristic of a vertical-cavity surface-emitting laser (VCSEL) [25]. The main feature of the VCSEL design is the regrown buried tunnel junction (BTJ) (see **Figure 6**), which accomplishes current confinement and wave guiding. The active region contains five heavily compressively strained InGaAs-quantum wells

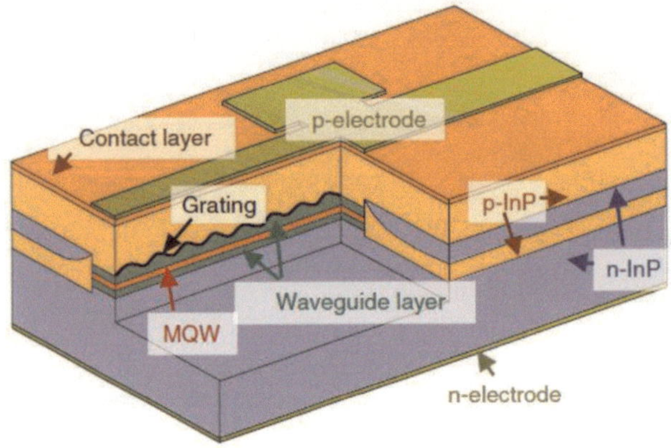

Figure 5.
Schematic of BH-DFB laser diode [23].

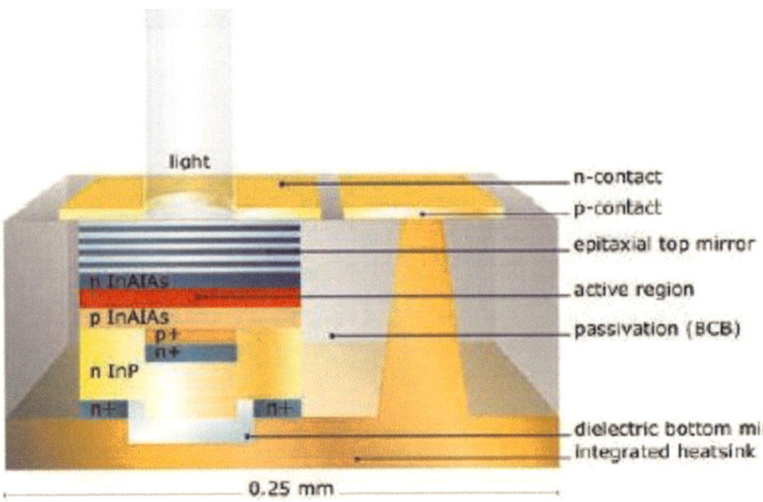

Figure 6.
Schematic of VCSEL laser diode [26].

separated by tensile strained barriers of InGaAIAs. Current is injected through a contact pad on the epitaxial mirror and the gold heat sink via the n + doped contact layers [25]. The n-doped epitaxial mirror reflectivity of the front mirror of 99.4% and back mirror reflectivity was 99.9%. The dielectric DBR is combined with an integrated electroplated Au-heat sink and a buried tunnel junction. VCSEL's emitting in the 2 μm region have been demonstrated by the Technical University Munich and commercialised by VERTILAS in the InP material system [25] however the ex-facet power levels from this technology is limited to <1 mW. Due to the very short laser cavity spectral linewidths from VCSEL devices are typically >20 MHz [26], and show high tuning rates of 0.67 nm/mA and 1.5 nm/°C.

2.2.3 InP-discrete mode laser diode

Discrete mode (DM) technology is Eblana Photonics proprietary method of manufacturing single mode lasers. Single wavelength operation in DM lasers is achieved by introducing index perturbations in the form of etched features

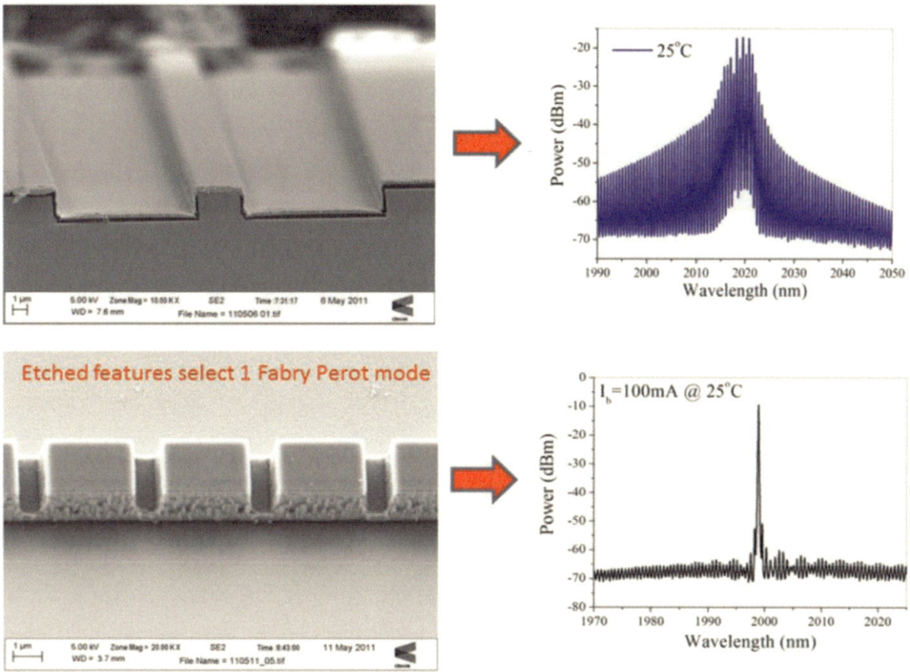

Figure 7.
Electron micrograph of Fabry Perot laser cavity (top left) and DM laser, fabricated by etching slots into laser ridge (bottom left) along with spectral characteristics of typical mid-IR laser with and without etched features (top and bottom right respectively).

positioned at a number of sites distributed along the ridge waveguide laser cavity as shown in **Figure 7**. Eblana has recently fabricated single mode lasers in the 2 μm spectral region [13]. The spectral linewidths for DM laser emitting at 2.0 μm is 1 MHz. The ex-facet power of 5 mW was measured and the wavelength shift with current and temperature is reported to be 0.0025 nm/mA and 0.125 nm/°C [13].

3. Mid-IR DM laser

3.1 Design of InP ridge waveguide FP lasers

A typical InP based Type-I laser structure for an emission wavelength of 1.6–2.1 μm is shown in **Figure 8** [14]. For most of the laser structures reported on in Section 3, two to three compressively strained QWs are used as the active region with a width of 10 nm. They are separated by 15–30 nm thick $In_xGa_{1-x}As$ barrier layers which are either lattice matched to the InP substrate or tensile strained in order to reduce the average strain in the active region. The active region is sandwiched between two 200 nm-thick InGaAsP separate confinement guide layers with a bandgap wavelength of λ_g = 1.3 μm which also acts as an etch stop layer for ridge waveguide definition. A 1800 nm thick p-InP layer is grown on top of the separate confinement heterostructure followed by a 200 nm thick highly p-doped $In_xGa_{1-x}As$ contact layer [3].

For the results shown below, four laser structures with varying In compositions were grown on 75 mm diameter n-type (100)-InP substrates in a metal-organic vapour-phase epitaxy reactor at low pressure. The overlapped photoluminescence spectra for the four wafers are shown in **Figure 9** measured at room temperature. The sharp peaks indicate high material quality with low defects.

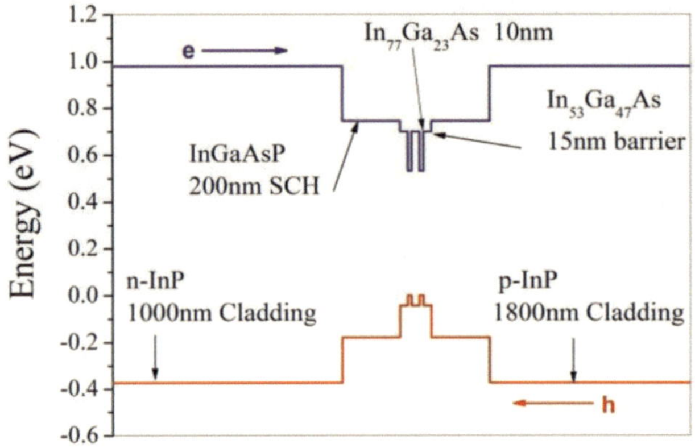

Figure 8.
Direct bandgap profile of an InP-based type-I laser structure with an emission wavelength of 2.1 μm [14].

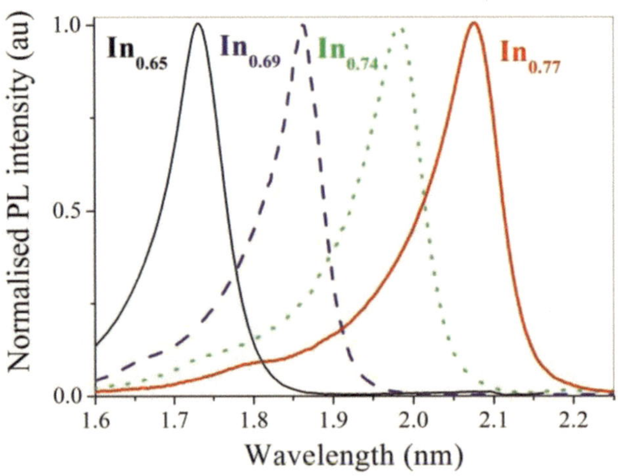

Figure 9.
Overlapped measured photoluminescence spectra at 25°C for four wafers with varying In composition.

The optical waveguiding properties of the ridge waveguide structure was determined by carrying out 2-D numerical simulations and one example for the 2 μm wafer (In = 0.74) is shown in **Figure 10**. The modal analysis of the structure showed that no higher order lateral modes are supported when the ridge width is 2 μm and that the calculated effective index for the waveguide is 3.2. This effective index is used to calculate the grating pattern spacing required to give single mode emission at the target wavelength which will be described below [3].

3.1.1 FP laser fabrication

In order to make good single longitudinal mode lasers, the ability to fabricate uniform, highly reliable FP lasers is essential and complete wafers of FP lasers were processed for material evaluation. For the laser to operate in a single lateral mode, control of the ridge width is critical. Simulations showed that a 2 μm wide ridge results in a stable transverse mode. The waveguide was realised using inductive coupled plasma dry etching, the dry etch chemistry used was Cl/N_2 followed by a short wet-etch to remove surface roughness. Electrical contacting was achieved using conventional metals (Ti/Pt/Au) and SiO_2 as an insulator for contact definition

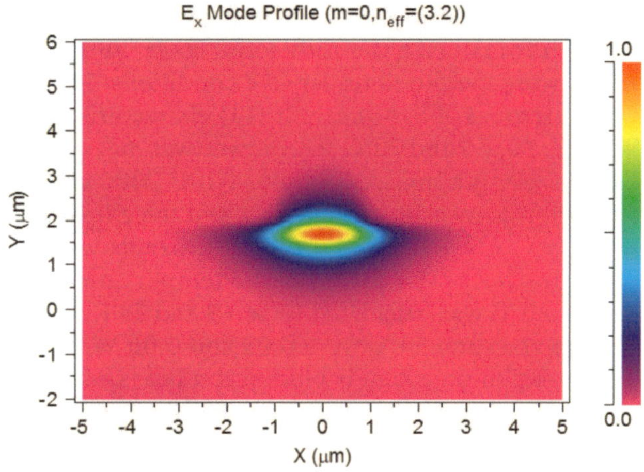

Figure 10.
2D simulation of a ridge waveguide laser diode.

on the heavily doped ($p \sim 2 \times 10^{19}$ cm^{-3}) p$^+$-InGaAs capped layer. A scanning electron microscope (SEM) image of the fabricated ridge waveguide is depicted in **Figure 11a**. Finally the wafers were thinned to 150 μm by mechanical polishing and the n-metal electrode applied [3]. A completed wafer is shown in **Figure 11b**. Subsequently bars were cleaved into 900 μm cavity lengths and the front and back facets coated 20 and 95% respectively.

3.1.2 Mid-IR FP laser characterization

To evaluate the material quality before fabricating the DM lasers, broad area and FP lasers with varying cavity length were fabricated and characterised on the four wafers, only the results from the 2 μm wafer are presented in this section. As-cleaved broad area lasers with 50 μm wide ridge widths and varying cavity lengths were analysed under pulsed conditions (1 μs pulses with 0.1% duty cycle) to evaluate the material quality. The threshold current density (J_{th}) for the 1000 μm cavity length was 358 A/cm^2 (~119 A/cm^2/QW) indicating good material quality. The slope efficiency variation with cavity length allowed the internal quantum efficiency (n_i) and the internal optical loss (n_i) to be estimated to be 80% and 8 cm^{-1} respectively.

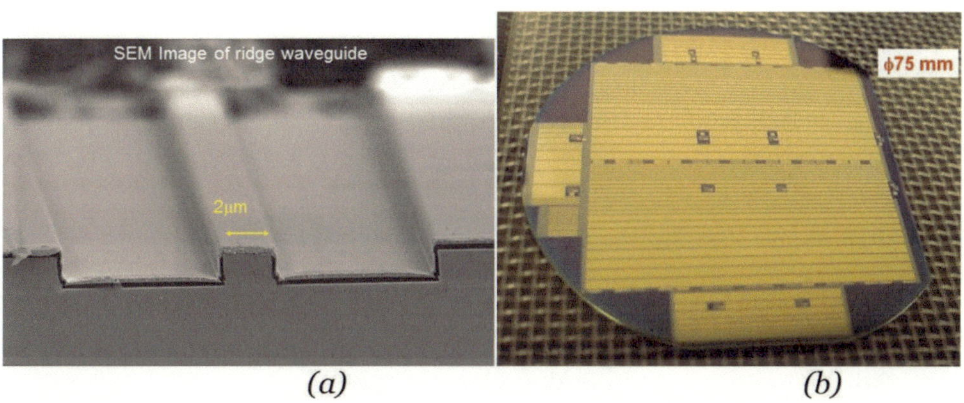

Figure 11.
(a) SEM image of a ridge waveguide FP laser diode. (b) Picture of a processed 3-inch InP wafer.

A 600 μm long FP ridge waveguide laser was packaged in a fiberised 14-pin butterfly module which contained a thermoelectric cooler and thermistor and the optical characteristics were measured under CW conditions. **Figure 12** shows the overlapped CW measurement of light-current (LI) characteristics measured at chip temperatures 10, 25, 45, 50, 60 and 70°C. The power was measured with a large area (Ø3 mm) extended wavelength InGaAs detector (GPD 3000). The light coupling efficiency into the fibre was measured to be 60% and the chip ex-facet power was >20 mW at 200 mA, 25°C. The extracted threshold currents were 18, 30 and 58 mA at 25, 50 and 70°C respectively. The characteristic temperature of threshold current between 25–50 and 25–70°C was calculated to be 49 and 38 K. The measured slope efficiencies in the fibre at 25 and 50°C were 0.08 and 0.06 W/A respectively [3].

In **Figure 13** the emission spectrum is plotted at a heat sink temperature of 25°C and bias current of 80 mA using a Yokogawa (*AQ6375*) long wavelength optical spectrum analyser. In **Figure 14** we overlap the measured optical spectra over a

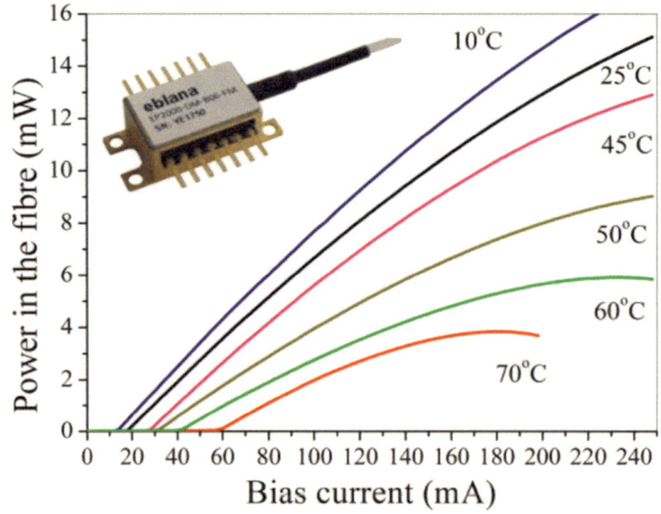

Figure 12.
Overlapped light-current curves in the temperature range 10–70°C for the 2 μm FP laser.

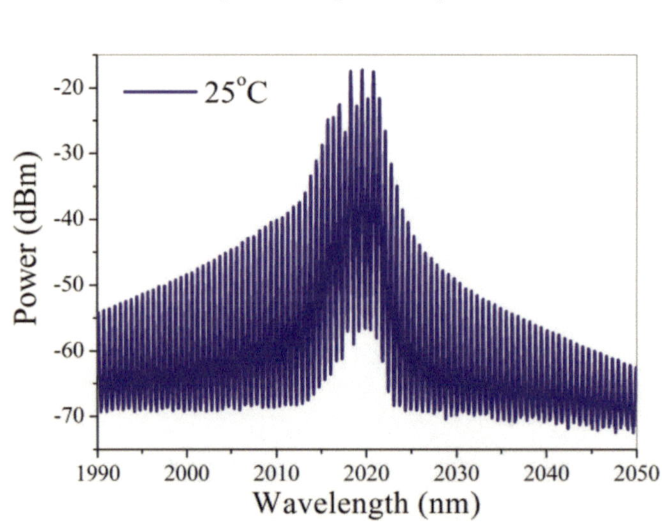

Figure 13.
Emission spectrum at a bias current of 80 mA and heat sink temperature of 25°C.

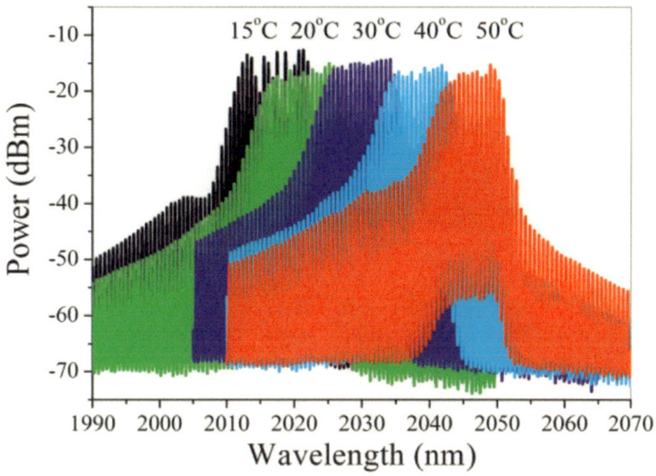

Figure 14.
Overlapped emission spectra at a bias current of 80 mA in the temperature range 25–50°C.

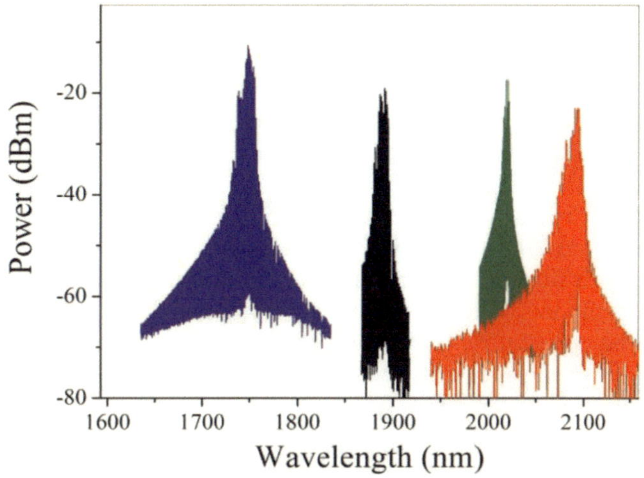

Figure 15.
Overlapped FP emission spectra at bias currents of 80 mA and heat sink temperature of 25°C for the four InP wafers.

temperature range from 15 to 50°C, the centre lasing wavelength shows a linear dependence with temperature with a tuning rate $\Delta\lambda/\Delta T$ of ~0.83 nm/°C, which is consistent with that expected due to the temperature-induced change in the refractive index.

To demonstrate the wide wavelength coverage from 1.65 to 2.15 μm ~400 nm we overlap the FP emission spectrum for the four wafers as shown in **Figure 15**.

3.2 Design of DM laser diodes

Single wavelength operation in DM laser diodes is achieved by introducing index perturbations in the form of shallow-etched features, or slots, positioned at a number of sites distributed along the ridge waveguide as shown in **Figure 16a** [27–32]. The slots are realized using ICP dry etching, with a typical depth in the region of 1.5–2 μm and a width of ~1 μm. The slots are relatively shallow and are not etched into the active (wave guiding) region, however, they will still interact with the

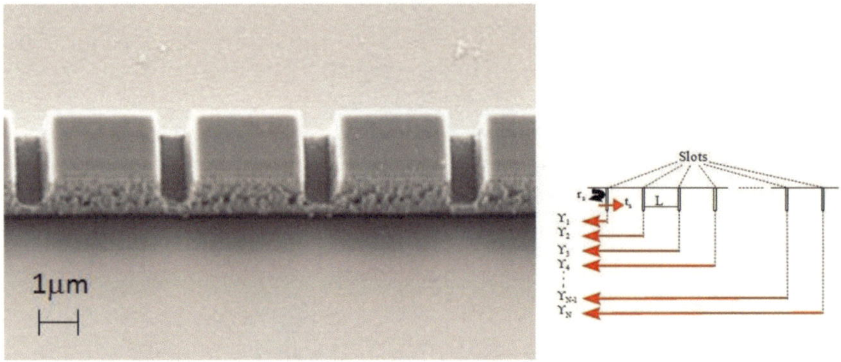

(a) (b)

Figure 16.
(a) SEM image of 2 μm wide ridge waveguide with etched grating. The slot width is 1 μm and the spacing between the slots L is 4 μm in this example. (b) Illustration of slot reflection and transmission for N slots.

mode's electric field as the mode profile is not fully confined to the active region and will expand into the surrounding cladding regions. This interaction results in a proportion of the propagating light being reflected at the boundaries between the perturbed and the unperturbed sections. In effect the slots act as reflection centres and through suitable positioning the slots manipulate the mirror loss spectrum of an FP laser so that the mirror loss of a specified mode is reduced below that of the other cavity modes [27–32]. Using a simplified model, developed in [33, 34], a slot can be described as a one dimensional discontinuity inserted into the cavity; as most of the reflection comes from the front of the slot interface. **Figure 16b** shows a schematic of a laser cavity with slots introduced into the cavity; where r_s is the slot reflectivity, t_s is the slot transmission, N is the number of slots, L is the distance between the slots, and Υ_i is the reflectivity in a section of the cavity where i is the slot number. The reflectivity from the first slot, Υ_1 is given by r_s. The introduction of a slot into the waveguide changes its effective refractive index, so that it differs slightly from the segments of the waveguide without slots. The reflectivity from the waveguide to slot interface can be approximated using Eq. (1):

$$r_s \approx abs\left(\frac{n_2 - n_1}{n_2 + n_1}\right) \tag{1}$$

where n_1 is the effective refractive index of the waveguide and n_2 is the effective refractive index of the waveguide with a slot.

Assuming no loss from the slot Eq. (2),

$$t_s = 1 - r_s \tag{2}$$

Υ_2 is the reflectivity from the second slot and is given by Eq. (3)

$$\Upsilon_2 = r_s t_s^2 \exp\left(-2i\beta L\right) \tag{3}$$

where β is the complex propagation constant, and the term t_s is squared to take account of forward and backward travelling waves. The exponential term describes the medium in which the light travels, and a factor of two is used again to take account of forward and backward travelling waves. The complex propagation constant takes account of the gain and loss in the transmission medium and is defined in terms of Eq. (4)

$$\beta = \beta_{re} + i\beta_i = \frac{2\pi n}{\lambda} + i\frac{g - \alpha_i}{2} \tag{4}$$

where n is the refractive index, λ is the wavelength, g is the optical gain and α_i is the internal cavity loss. The reflectivities of the third and fourth slots are given by Eq. (5):

$$\Upsilon_3 = r_s t_s^4 \exp(-4i\beta L) \tag{5}$$

and Eq. (6),

$$\Upsilon_4 = r_s t_s^6 \exp(-6i\beta L) \tag{6}$$

respectively; therefore, the reflectivity obtained from four slots is given by Eq. (7):

$$\begin{aligned}\Upsilon_{\text{total}} &= \Upsilon_1 + \Upsilon_2 + \Upsilon_3 + \Upsilon_4 \\ &= r_s + r_s t_s^2 \exp(-2i\beta L) + r_s t_s^4 \exp(-4i\beta L) + r_s t_s^6 \exp(-6i\beta L)\end{aligned} \tag{7}$$

By letting Eq. (8)

$$X = t_s^2 \exp(-2i\beta L) \tag{8}$$

the total reflectivity from N slots can be expressed by Eq. (9) the following series.

$$\Upsilon_{\text{total}} = r_s\left(1 + X + X^2 + X^3 + \ldots\ldots + X^{N-1}\right) \tag{9}$$

Which in terms of known variables can be described as Eq. (10)

$$\Upsilon_{\text{total}} = r_s\left[\frac{1 - \left(t_s^2 \exp(-2i\beta L)\right)^N}{1 - t_s^2 \exp(-2i\beta L)}\right] \tag{10}$$

The power reflection is related to the reflection amplitude by Eq. (11),

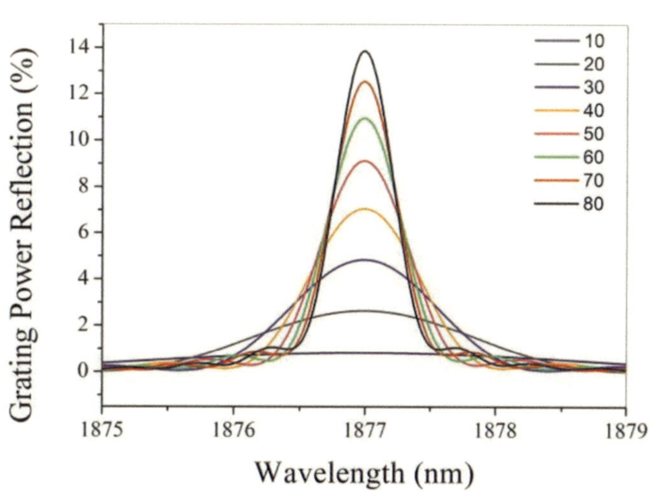

Figure 17.
Simulated power reflection spectrum as a function of slot number.

$$R = abs\left(\Upsilon_{total}^{2}\right) \quad (11)$$

Using this model the power reflection versus etched feature number at a wavelength of 1887 nm was simulated and shown in **Figure 17**.

For a 600 μm long laser cavity at about 60 slots the peak reflectivity begins to saturate and the FWHM is about 0.9 nm which is equivalent to the FP mode spacing. So this is the optimum number of slots for this cavity length.

3.2.1 DM laser fabrication

The fabrication of the DM laser is exactly the same as the FP laser outlined in Section 3.1.2 width the addition of one extra dry etching step to etch the grating, the dry etch chemistry used again was Cl/N2 and was followed by a short wet-etch to remove surface roughness from the grating. A schematic of the etched features is shown in **Figure 18**.

3.2.2 DM laser diodes emitting at 1.87 μm

Low cost single mode semiconductor laser diodes emitting at 1.87 μm are highly desirable as light sources for trace gas sensing of H_2O. The measurement of H_2O is important in many industrial applications, for example, continuous emission monitoring in combustion processes where the vapour concentration can be related to performance parameters, such as, efficiency of combustion and heat release. In this section we present data on DM laser diodes operating around λ = 1.877 μm. Fabricated DM lasers exhibit continuous wave (CW) mode hop free operation in the temperature range from 15 to 55°C with emission wavelengths centred at 1.87 μm at 25°C, and ex-facet optical output power >3 mW at 25°C.

A ridge waveguide 600 μm in length DM laser diode was die bonded to an aluminium nitride submount and the optical characteristics were measured under CW conditions. **Figure 19** plots the overlapped CW measurement of ex-facet LI characteristics measured at chip temperatures of 15, 25, 35, 45 and 55°C, with the same power and wavelength measurement setup as mentioned in the previous section. The extracted threshold currents were 12, 15, 18, 22 and 29 mA at 15, 25, 35, 45 and 55°C, respectively. The measured slope efficiencies at 15 and 55°C were 0.038 and 0.022 W/A, respectively.

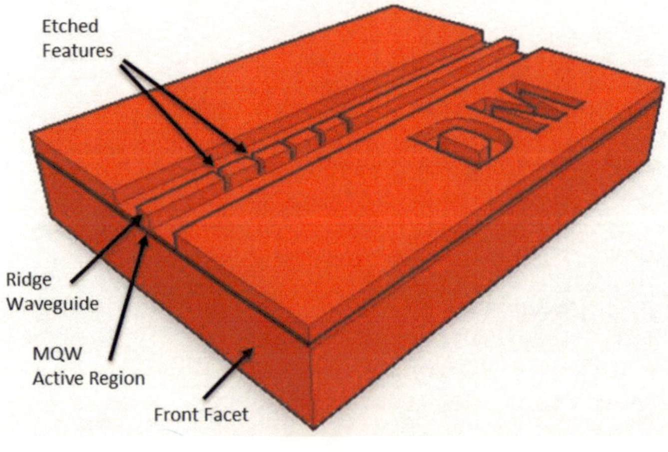

Figure 18.
Schematic view of the InP DM laser.

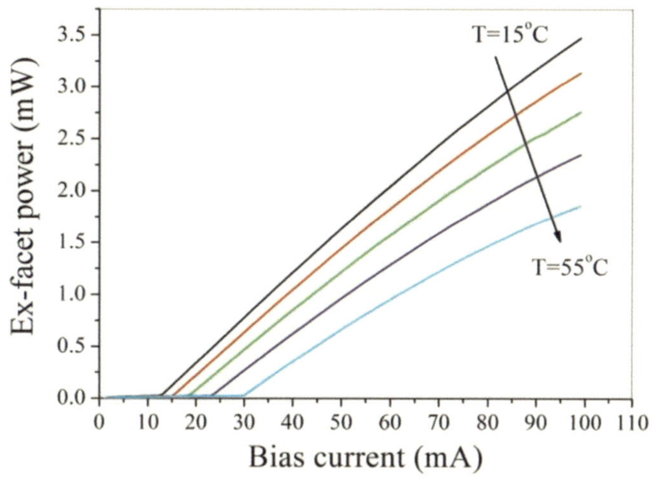

Figure 19.
Overlapped CW LI curves as a function of heat sink temperature.

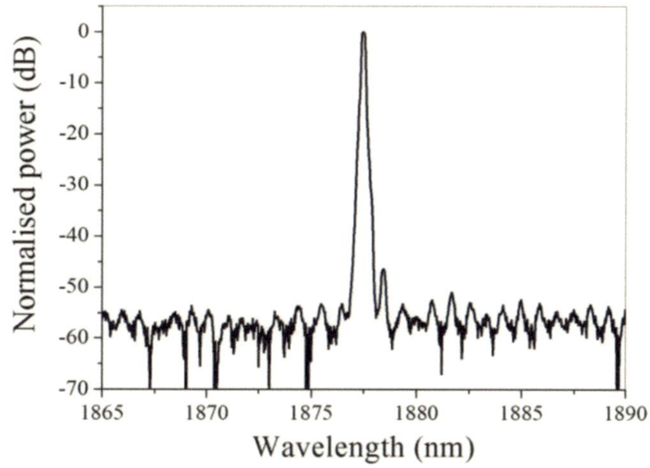

Figure 20.
Normalised optical emission spectrum at a bias current of 100 mA and heat sink temperature of 25°C with SMSR >45 dB demonstrated.

In **Figure 20** the emission spectrum of the DM laser diode is measured at a heat sink temperature of 25°C and bias current of 100 mA. A peak wavelength of 1877 nm is demonstrated, with a side mode suppression ratio (SMSR) of ~45 dB achieved, in excellent agreement with simulated values. In **Figure 21a** we plot the peak wavelength versus bias current as a function of laser submount temperature. The peak wavelength tunes linearly with bias current at a tuning rate of ~0.017 nm/mA. In **Figure 21b** the optical emission spectrum over a wide temperature range, from −5 to 55°C, is plotted. The single mode peak lasing wavelength shows a linear dependence with current and temperature, with a tuning rate of $\Delta\lambda/\Delta I$ ~0.017 nm/mA and $\Delta\lambda/\Delta T$ of 0.113 nm/°C, consistent with that expected due to the temperature-induced change in the refractive index [29].

Wide gain bandwidth is demonstrated in **Figure 22** where the DM peak wavelength is varied by changing the grating period wavelength tunability of 120 nm is achieved. This makes this material promising for widely-tunable mid-infrared single-mode devices such as DM arrays, external cavity lasers and sampled gratings. Eblana's DM technology has been used to demonstrate single frequency lasers with

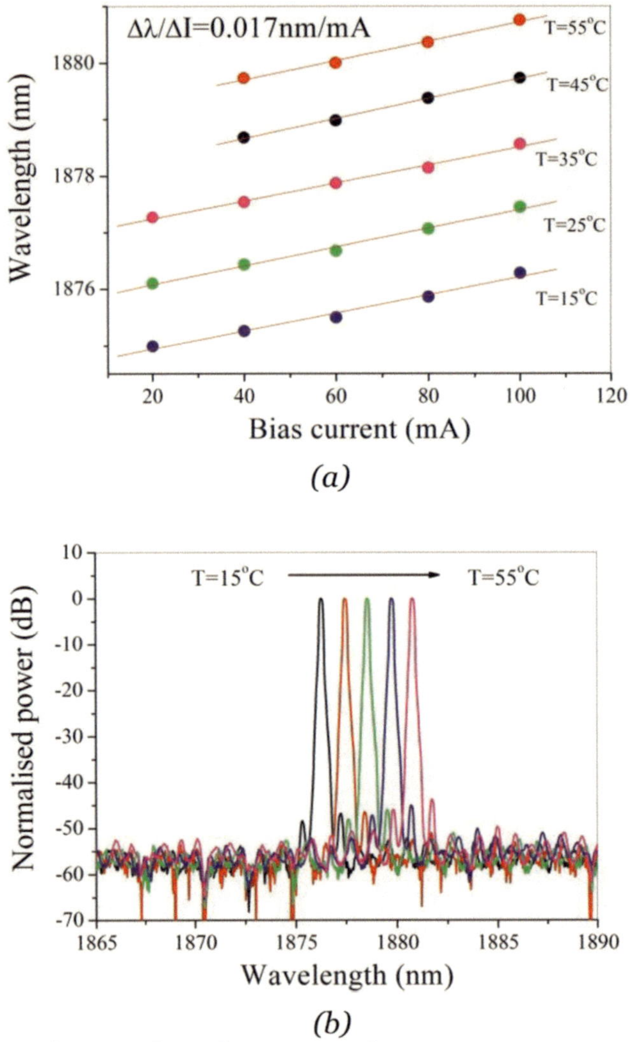

Figure 21.
(a) Peak wavelength versus bias current over temperature. (b) Overlapped optical emission spectra versus heat sink temperature at a fixed laser bias current of 100 mA.

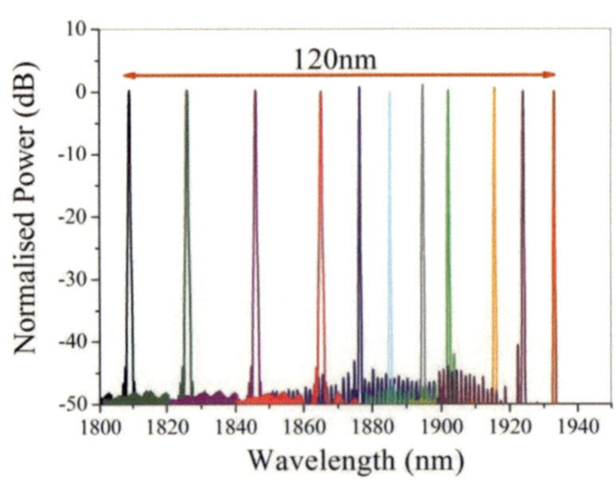

Figure 22.
Overlapped single frequency spectra of 10 different DM lasers demonstrating 120 nm wide gain for the 1.8 μm wafer.

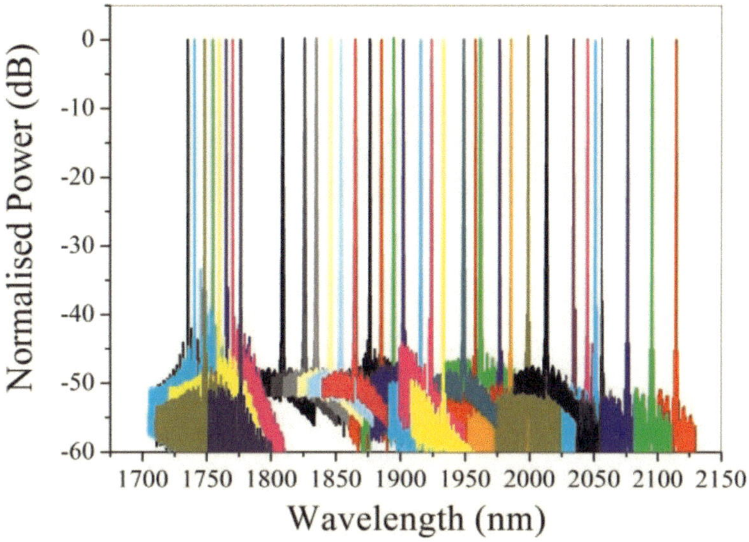

Figure 23.
Overlapped single frequency spectra of different DM lasers based on four InGaAs quantum wells wafers spanning the 1.7–2.15 μm wavelength region.

high side mode suppression ratios spanning a wide wavelength range from 1.75 to 2.1 μm by using the appropriate InGaAs quantum well composition and thicknesses as shown in **Figure 23**.

The wavelength and tuning results demonstrated by DM lasers in this chapter show that the devices operate at a single wavelength with SMSR >40 dB robustly over current and temperature variations and, furthermore, that these devices are highly suitable as low cost sources for TDLAS and other sensor applications. A further important feature of the DM laser diode is that its fabrication is far less complex than that of a distributed feedback laser diode resulting in a significant cost advantage.

4. Conclusion

In this chapter an overview in the current state of the art in mid-IR single mode lasers was presented and a low cost laser technology platform based on the InP material system for the manufacture of single mode laser diodes in the near to mid-IR wavelength range introduced. These DM lasers, are ridge waveguide Fabry Perot lasers whose emission spectra have been modified to produce a single mode operation. This modification is achieved during wafer processing by etching surface features into the ridge of an otherwise conventional ridge waveguide laser diode structure and therefore avoiding the need for grating overgrowth [35].

The basic principles underlying DM laser operation and their fabrication have been described. Basic static characteristics of the devices have been discussed and operation in the 1.6–2.1 μm spectral window demonstrated. The results demonstrate that InP-based light sources provide a promising concept for mid-infrared semiconductor diode lasers and are suitable for a range of sensor and other applications.

InP based diode lasers covering the 1.8–2.1 μm range have already reached a considerable level of maturity, as evidenced by the low threshold currents and good spectral performance as highlighted in Section 3. Thus current R&D focuses on the optimization of these lasers towards longer wavelengths >2.3 μm where specific application are in high demand.

Acknowledgements

This work was supported in part by the European project MODEGAP (FP/2007-2013 under grant agreement 258033) and by the European Space Agency (ESA) under contract reference ESTEC/ITT AO/1 7204/12/NL/NA.

Author details

Richard Phelan*, Diarmuid Byrne, John O'Carroll, Michael Gleeson, Marta Nawrocka, Rob Lennox, Kevin Carney, Chris Herbert, Jim Somers and Brian Kelly
Eblana Photonics Ltd., Dun Laoghaire, Ireland

*Address all correspondence to: richard.phelan@eblanaphotonics.com

References

[1] Rothman LS, Gordon IE, Babikov Y, Barbe A, Benner DC, Bernath PF, et al. The HITRAN2012 molecular spectroscopic database. Journal of Quantitative Spectroscopy and Radiation Transfer. 2013;**130**:4-50

[2] Stephan S, Frederic D, Christian AM. Novel InP- and GaSb-based light sources for the near to far infrared. Semiconductor Science and Technology. 2016;**31**:113005

[3] Available from: http://www.modegap.eu/

[4] Poletti F, Wheeler NV, Petrovich MN, Baddela N, Numkam-Fokoua E, Hayes JR, et al. Towards high-capacity fibre-optic communications at the speed of light in vacuum. Nature Photonics. 2013;7:279-284

[5] Mac Suibhne N, Li Z, Baeuerle B, Zhao J, Wooler J, Alam S, et al. WDM transmission at 2 μm over low-loss hollow core photonic bandgap fiber. In: Optical Fiber Conference; Los Angeles, USA; 2013. OW1I.6

[6] Mac Suibhne N, Li Z, Baeuerle B, Zhao J, Wooler J, Alam S, et al. Wavelength division multiplexing at 2 μm. In: European conference on Optical communication; Amsterdam, Netherlands; 2012. TH.3.A.3

[7] Olesberg JT. Noninvasive blood glucose monitoring in the 2.0 μm spectral range. In: 2001 IEEE/LEOS Conference Proceedings; Vol. 2; pp. 529

[8] Lauer C, Ortsiefer M, Shau R, Rosskopf J, Bohm G, Ronneberg E, et al. 80°C continuous-wave operation of 2.01 μm wavelength InGaAlAs-InP vertical-cavity surface-emitting lasers. IEEE Photonics Technology Letters. 2004;**16**:2209-2211

[9] Major JS Jr, Nam DW, Osinski JS, Welch DF. High-power 2.0 μm InGaAsP laser diodes. IEEE Photonics Technology Letters. 1993;**5**(6):594-596

[10] Serries D, Peter M, Kiefer R, Winkler K, Wagner J. Improved performance of 2-μm GaInAs strained quantum-well lasers on InP by increasing carrier confinement. IEEE Photonics Technology Letters. 2001; **13**(5):412-414

[11] Mitsuhara M, Ogasawara M, Oishi M, Sugiura H. Metalorganic molecular-beam-epitaxy-grown In Ga As/InGaAs multiple quantum well lasers emitting at 2.07 μm wavelength. Applied Physics Letters. 1998;**72**:3106-3108

[12] Forouhar S, Keo S, Larsson A, Ksendzov A, Temkin H. Low threshold continuous operation of InGaAs/InGaAsP quantum well lasers at 2.0 μm. Electronics Letters. 1993;**29**:574-576

[13] Phelan R, O'Carroll J, Byrne D, Herbert C, Somers J, Kelly B. In0.75Ga0.25As/InP multiple quantum-well discrete-mode laser diode emitting at 2 μm. IEEE Photonics Technology Letters. 2012;**24**(8):652-654

[14] Phelan R, Gleeson M, Byrne D, O'carroll J, Nawrocka M, Carney K, et al. High power narrow linewidth discrete mode laser diode integrated with a curved semiconductor optical amplifier emitting at 2051 nm. Applied Optics. 2018;**57**(22):E1-E5

[15] Sato T, Mitsuhara M, Nunoya N, Fujisawa T, Kasaya K, Kano F, et al. 2.33 mm wavelength distributed feedback lasers with InAs-InGaAs multiple-quantum wells on InP substrates. IEEE Photonics Technology Letters. 2008;**20**(12):1045-1047

[16] Rattunde M. III-Sb-Based Type-I QW Diode Lasers. In: Anthony K, editor. Mid-infrared Semiconductor

Optoelectronics. Springer Series in Optical Sciences; 2006. pp. 131-157

[17] Sprengel S, Grasse C, Vizbaras K, Gruendl T, Amann M-C. Up to 3 μm light emission on InP substrate using GaInAs/GaAsSb type-II quantum wells. Applied Physics Letters. 2011;**99**:221109

[18] Forouhar S et al. High-power laterally coupled distributed-feedback GaSb-based diode lasers at 2 μm wavelength. Applied Physics Letters. 2012;**100**:031107

[19] Bagheri M, Frez C, Kelly B, Gupta JA, Forouhar S. High output power, fibre-coupled distributed feedback lasers operating near 2.05 μm wavelength range. Electronics Letters. 2013;**49**(24):1552-1553

[20] Bleuel T et al. 2 μm GaInSb-AlGaAsSb distributed-feedback lasers. IEEE Photonics Technology Letters. 2001;**13**(6):553-555

[21] Barbero J, López D, Esquivias I, Tijero JMG, Fischer M, Roessner K, et al. Evaluation of 2.1 μm DFB lasers for space applications. In: Proceedings Volume 10565, International Conference on Space Optics—ICSO 2010;**2017**:1-4. 1056523

[22] Mitsuhara M et al. 2.05 μm wavelength InGaAs-InGaAs distributed-feedback multiquantum-well lasers with 10-mW output power. IEEE Photonics Technology Letters. 1999;**11**(1):33-35

[23] Sato T et al. 2.33 μm-wavelength distributed feedback lasers with InAs-InGaAs multiple-quantum wells on InP substrates. IEEE Photonics Technology Letters. 2008;**20**(12):1045-1047

[24] Takeshita T et al. Long-term degradation behaviour of 2.3-μm wavelength highly strained Incas/InP MQW-DFB lasers with a p−/n-InP buried heterostructure. IEEE Transactions on Electron Devices. 2012;**59**(4):1056-1062

[25] Lauer C, Ortsiefer M, Shau R, Rosskopf J, Bohm G, Maute M, et al. Electrically pumped room temperature CW-VCSELs with emission wavelength of 2 μm. Electronics Letters. 2003;**39**(1): 57-58

[26] Shau R et al. Linewidth of InP-based 1.55 μm VCSELs with buried tunnel junction. Electronics Letters. 2003;**39**(24):1728-1729

[27] Phelan R, Kelly B, O'Carroll J, Herbert C, Duke A, O'Gorman J. −40°C < T < 95°C mode-hop free operation of an uncooled AlGaInAs-MQW discrete-mode laser diode with emission at λ = 1.3 μm. IEE Electronics Letters. 2009;**45**(1):43-45

[28] Phelan R, Slight TJ, Kelly B, O'Carroll J, McKee A, Revin DG, et al. O'Gorman 'Room temperature operation of discrete mode InGaAs/AlAsSb quantum cascade laser with emission at λ = 1.3 μm' IEEE Photonics Technology Letters Year: 2010, Volume: 22, Issue:17 Pages: 1273-1275

[29] Phelan R, Gleeson M, Byrne D, O'Carroll J, Long P, Maigyte L, et al. InGaP/AlGaInP quantum well discrete mode laser diode emitting at 689 nm. IEEE Photonics Technology Letters. 2018;**30**(3):235-237

[30] Patchell J, Jones D, Kelly B, O'Gorman J. Specifying the wavelength and temperature tuning range of a Fabry-Perot laser containing refractive index perturbations. In: Proceedings of SPIE; vol. 5825; April. 2005. pp. 1-13

[31] O'Carroll J. Novel optical transmitters for high speed optical networks [thesis]. Dublin City University; 2013

[32] Byrne DC. An experimental investigation of tunable laser diodes based on multiple etched slots [thesis]. Dublin: Trinity College; 2010

[33] Byrne DC, Engelstaedter JP, Guo WH, Qiao YL, Corbett B, Roycroft B, et al. Discretely tunable semiconductor lasers suitable for photonic integration. IEEE Journal of Selected Topics in Quantum Electronics. 2009;**15**(3): 482-487

[34] Lu Q, Guo W, Phelan R, Byrne D, Donegan JF, Lambkin P, et al. Analysis of slot characteristics in slotted single-mode semiconductor lasers using the 2-D scattering matrix method. IEEE Photonics Technology Letters. 2006; **18**(24):2605-2607

[35] Herbert C, Jones D, Kaszubowska-Anandarajah A, Kelly B, Rensing M, O'Carroll J, et al. Discrete mode lasers for communication applications. IET Optoelectronics. 2009;**3**(1):1-17

Chapter 3

Optoelectronic Key Elements for Polymeric Fiber Transmission Systems

Ulrich H.P. Fischer-Hirchert and Mladen Joncic

Abstract

In short-range communication 1 mm PMMA SI-POF established itself as a reasonable alternative to the traditional data communication media such as glass fibers, copper cables, and wireless systems. Due to multiple advantages such as a large core diameter, tolerance to fiber facet damages, and low installation costs, the SI-POF is already applied in industrial automation, automotive industry, and in-house/office networks. To experimentally demonstrate the feasibility and potential of a high-speed POF WDM concept, a four-channel data transmission setup was realized. A four-legged multiplexing POF bundle was developed to combine the signals from four visible laser diodes onto SI-POF link. For the separation of wavelength channels, the interference filter-based demultiplexer with two-stage configuration was used. It was shown that POF WDM with lower channel rates and simple transmission technique (NRZ + FFE) could provide aggregate bit rates comparable to those achieved with the single-wavelength systems that used advanced modulation formats (DMT or PAM + DFE) and required significant signal processing. In addition, the 50 m SI-POF link at an aggregate bit rate of 7.8 Gb/s was demonstrated over 50 m SI-POF, respectively, at the BER = 10^{-3}.

Keywords: polymeric fiber transmission, WDM over POF, optical DMT modulation, optical PAM modulation

1. Introduction

Having its origin in the 1960s as well as the silica glass fiber, the polymer optical fiber (POF) stayed long in the shadow of the huge development and success of glass fiber communications. However, the advances in POF technology and the growing need for high-speed short-range communication networks make POF nowadays gain more and more importance. The key advantage of POF is a large core diameter. It makes POF tolerant to the fiber facet damages and relaxes the alignment tolerances, thus also reducing the installation costs. Furthermore, POF is pliable, durable, and inexpensive; offers small weight and short bend radius; allows easy installation, simple termination, and quick troubleshooting; and also provides the immunity to electromagnetic interference. Due to its diverse advantages, in short-range applications POF established itself as a reasonable alternative to the traditional data communication media such as glass fibers, copper cables, and wireless systems (see **Table 1**).

Transmission medium	Data rate	Distance	Safety	Cost	Handling	Installation	Total
Twisted pair cable	+	0	0	++	—	0	2+
Coaxial cable	0	0	0	+	0	0	1+
Glass fiber	++	++	++	—	—	—	1+
Polymer fiber	0	—	++	+	+	+	4+
Wireless	—	—	—	++	++	++	1+
Powerline	—	—	—	+	+	++	0

Table 1.
Comparison of different transmission media. Characteristics between very bad (−−) and particularly good (++) [1].

Today, POF is produced with different core materials, core diameters, and index profiles. A comprehensive overview on various POFs is given in [1]. Two major POF types are made of polymethyl methacrylate (PMMA) and perfluorinated (PF) materials. The parameters of the common PMMA and PF POFs are specified in the IEC Standard 60793-2-40, which defines eight different POF classes [2]. The PMMA POF is produced with both step-index (SI) and graded-index (GI) profile, whereas the PF POF offers only GI profile. The GI profile of the core ensures high modal bandwidth exceeding 1.5 GHz × 100 m for the PMMA POF and 300 MHz × 1 km for the PF POF. However, the implementation of the PMMA GI-POF is confined to 500–680 nm wavelength range due to the high optical attenuation at other wavelengths (>400 dB/km). In contrast, the PMMA SI-POF suffers from intermodal dispersion limiting the bandwidth-length product to around 50 MHz × 100 m but also provides several attenuation windows in the visible spectrum (400–700 nm). Due to its advantages over the other POF types such as technological maturity, ease and cost of production, and high numerical aperture (NA), the standard 1 mm PMMA SI-POF (POF class A4a.2 according to IEC 60793-2-40) is the best known and by far the most widely employed type of POF. This is also the fiber this work concentrates on.

In vehicles SI-POF displaces copper in the network structure of a passenger cabin for multimedia data services. The infotainment communication system known as Media Oriented System Transport (MOST) connects different multimedia components in the SI-POF-based ring topology [3], as illustrated in **Figure 1**. The current (third) version of the MOST system (MOST150) supports the data transfer at 150 Mb/s over link lengths of about 10 m.

Another sector where SI-POF displaces traditional communication media are short-range networks in houses and offices. As an in-house extension of a broadband access network (e.g., VDSL, HFC, FTTB), the typical application of POF technology is the delivery of triple-play services (combination of broadcasting, telecommunication, and the Internet) to the end user. The Fast Ethernet transceivers (100 Mb/s) and since 2013 also the Gigabit Ethernet transceivers (1 Gb/s) are available on the market enabling the transmission of broadband services over 50 m SI-POF. The Gigabit solutions from KD-POF employing the multilevel signaling and from Teleconnect based on the multicarrier modulation are accompanied by the technical standards ETSI TS 105175-1-2 [4] and ITU-T G.9960, Annex F [5], respectively.

The commercial communication systems with SI-POF use a single channel for data transmission. However, the transmission performances of SI-POF are impaired by strong intermodal dispersion and high optical attenuation. Stimulated by the growing bandwidth demands (e.g., 10–40–100 Gb/s Ethernet speed), various concepts to overcome the low-pass characteristic of SI-POF have been successfully demonstrated over the last few years. The simplest solutions utilized passive

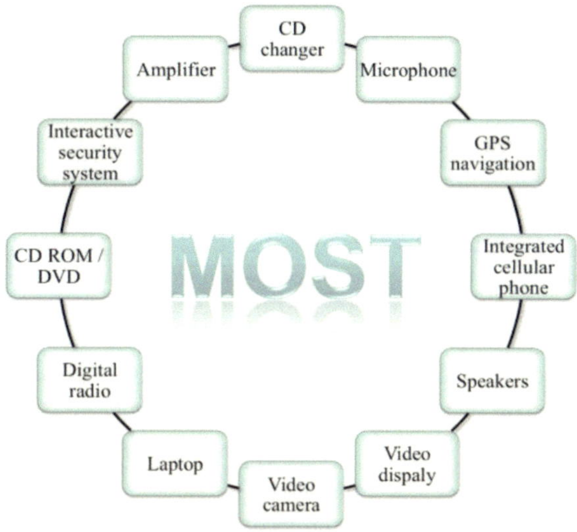

Figure 1.
SI-POF-based ring topology of a MOST system in a car.

equalization implemented as an analog high-pass filter that increased the electrical −3 dB bandwidth of a channel [6]. A major focus was also placed on the digital signal processing techniques, which were mostly implemented offline due to the lack of commercial components. Both the non-return-to-zero (NRZ) and the spectrally efficient multilevel signaling were combined with the digital receiver equalization to increase the data rates over SI-POF [7, 8]. The sophisticated spectrally efficient multicarrier modulation formats were also successfully implemented to combat the highly dispersive SI-POF channel [9, 10].

Complying with any of the hitherto developments, utilization of several optical carriers for parallel transmission of data channels over a single fiber represents another alternative to increase the transmission capacity of SI-POF. The technique is well-known as wavelength division multiplexing (WDM). The principle of WDM is shown in **Figure 2**. Since different wavelengths λ_1–λ_N do not interfere with each other in a linear medium, they can be used to simultaneously carry the data signals over a single fiber. Thereby, the capacity of a fiber, i.e., of an optical communication system, increases almost proportionally with the number of wavelength channels.

Two components are essential for WDM, a wavelength multiplexer and demultiplexer. The multiplexer combines the signals at different wavelengths, coming from different transmitters, onto a single fiber. On the opposite side of the optical link, the demultiplexer performs an inverse function, separating the wavelength channels to be detected by separate receivers.

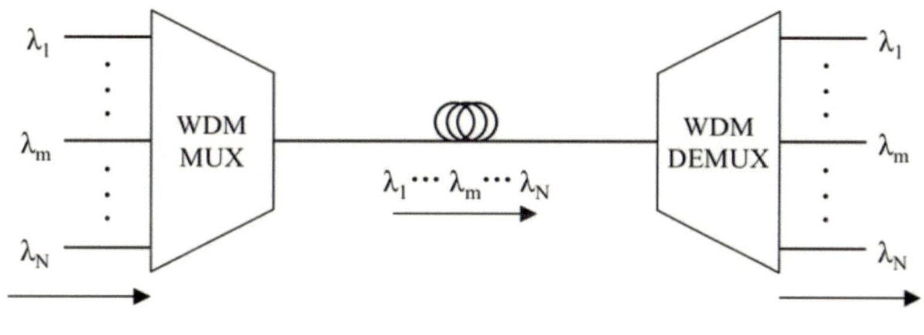

Figure 2.
Principle of WDM: MUX, multiplexer; DEMUX, demultiplexer.

The existing WDM components developed for single-mode glass fibers in the infrared region, such as Mach-Zehnder interferometers, arrayed waveguide gratings or fiber Bragg gratings, cannot be reused for a highly multimode SI-POF. On the other hand, the operating principles of demultiplexers based on thin-film interference filters and on a diffraction grating can be applied for POF. In spite of some other demultiplexing solutions (e.g., employing dispersion prisms), these two demultiplexing techniques have been recognized as the most promising for SI-POF. However, because of the difference in the operating wavelength range, fiber diameter, NA, etc., compared to the glass fibers, such demultiplexers must be newly designed for SI-POF communication. An overview of the state-of-the-art thin-film interference filter-SI-POF demultiplexers will be given. The first aim of this work is to further investigate experimentally these demultiplexing techniques for SI-POF. Accordingly, the aim of this work is to demonstrate experimentally high-speed POF WDM data transmission offering capacity increase compared to the single-channel systems.

2. Fundamentals of WDM for short-range communication over POF

2.1 Transmission properties of 1 mm PMMA SI-POF

The 1 mm PMMA SI-POF is the best known and by far the most widely employed type of POF. It is made of 980 μm diameter PMMA core surrounded by a thin cladding (10 μm) made of fluorinated polymer. The typical spectral attenuation of SI-POF is shown in **Figure 3**. The fiber supports operation in the visible spectrum from 400 to 700 nm. The lower wavelength bound is determined by the degradation of the PMMA compound with prolonged exposure to the ultraviolet (UV) wavelengths shown in [11] and in [12]. The attenuation value of around 400–450 dB/km, which still allows operation over shorter link lengths (<20 m), sets the upper wavelength bound.

Two intrinsic loss mechanisms contribute to the raise of attenuation at shorter and particularly UV wavelengths. The electronic transitions due to the absorption of light in the polymer compound cause absorption peaks in the UV region. However, their absorption tails extend through the visible spectrum affecting the POF attenuation [13]. The dependence of the attenuation coefficient of electronic transitions α_e [dB/km] on the wavelength for PMMA is given by [14]:

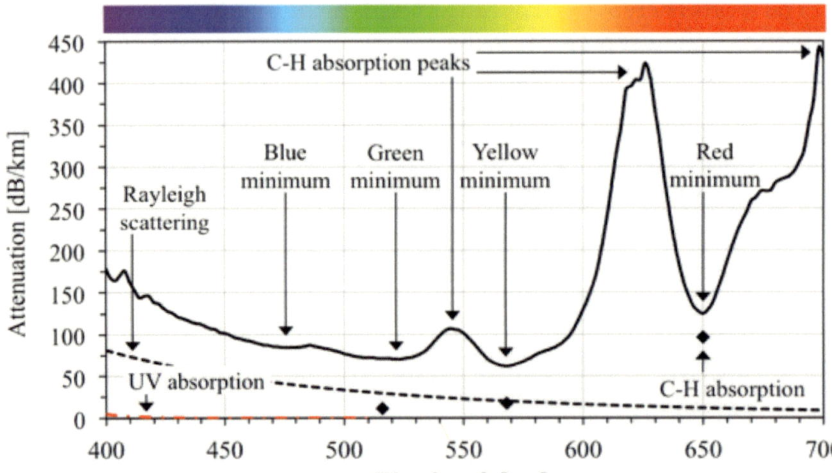

Figure 3.
Typical spectral attenuation of 1 mm PMMA SI-POF [13] with contributions of intrinsic loss mechanisms and with attenuation minima and maxima.

$$\alpha_e = 1.58 \cdot 10^{-12} exp\left(\frac{1.15 \cdot 10^4}{\lambda}\right). \tag{1}$$

The second loss mechanism is the Rayleigh scattering. It is caused by the structural irregularities in the polymer compound that are much smaller than the wavelength of light (order of one tenth of wavelength or less). The effect of scattering becomes more pronounced as the wavelength decreases since the scattering attenuation coefficient α_s [dB/km] is inversely proportional to the fourth power of the wavelength shown by Kaino [15]:

$$\alpha_s = 13 \cdot \left(\frac{633}{\lambda}\right)^4 \tag{2}$$

In the infrared region, the attenuation significantly increases due to the intrinsic absorption losses caused by vibrations of the molecular C-H bonds (total of eight per MMA monomer). The higher overtones of the C-H bond vibrations also extend in the visible spectrum. The seventh overtone at 549 nm and particularly the sixth and the fifth overtone at 627 and 736 nm, respectively, cause pronounced absorption peaks and wide absorption bands, predominantly determining the level of attenuation in the red spectral range shown by Emslie [16] and by Groh [17].

The contributions of the intrinsic loss mechanisms to the overall attenuation of SI-POF are also shown in **Figure 3**. The wavelength regions where the fiber exhibits low attenuation are called attenuation windows. The SI-POF has four attenuation windows. Those are blue, green, yellow, and red windows, with the absolute attenuation minimum of approx. 62 dB/km at around 568 nm (yellow window). The parameters of the attenuation windows are listed in **Table 2**.

The mean refractive index of SI-POF core material in the visible spectrum is n_{core}=1.492, whereas the refractive index of cladding is n_{clad}=1.412. Due to the big difference in refractive indices of core and cladding, the numerical aperture (NA)

$$NA = \sqrt{n_{core}^2 - n_{clad}^2} \tag{3}$$

has the value of 0.482 (usually rounded to 0.5). The corresponding maximum acceptance angle of the fiber is 30°. The large core radius a_{core} = 490 μm combined with the high NA results in the normalized frequency V

$$V = 2\pi\frac{a_{core}}{\lambda} \cdot NA \tag{4}$$

of 2698 at 550 nm, which is far above the limit V = 2.405 below which a fiber is in single-mode operation. The number of modes N_{mod} propagating through SI-POF can be approximated as

$$N_{mod} \approx V^2/2, \tag{5}$$

Attenuation window	Blue	Green	Yellow	Red
Attenuation minimum [dB/km]	85	70	62	125
Wavelength of the attenuation minimum [nm]	476	522	568	650
Approximate 3 dB width of the window [nm]	19	24	8	4

Table 2.
Attenuation windows of SI-POF (based on the attenuation curve from **Figure 2***).*

corresponding to 3.64 million modes at 550 nm. Due to the significant path difference between lower- and higher-order modes, propagating respectively at smaller and larger angles relative to the optical axis, the strong intermodal dispersion is inherent to SI-POF. In the time domain, it is manifested as pulse broadening, thus introducing the inter-symbol interference (ISI). In the frequency domain, the intermodal dispersion results in a low-pass frequency response, constraining the bandwidth-length product of SI-POF to around 50 MHz × 100 m shown by Ziemann et al. [18].

3. Demultiplexing employing thin-film interference filters

The technology based on thin-film interference filters is mature and one of the most commonly applied technologies for realization of WDM demultiplexers in single-mode glass fiber communication. The demultiplexers for coarse WDM applications cascade the interference filters to provide up to 16 flattop channels between 1271 and 1611 nm, with 20 nm minimum channel spacing [19]. The typical parameters of commercial 4-, 8-, and 16-channel demultiplexers with IL <1.6, 2.7, and 3.7 dB, respectively, can be found in [20]. The thin-film filter-based demultiplexers for dense WDM applications are commercially available with up to 40 channels in 1550 nm region and <8 dB IL. Instead of simply cascading the filters, those devices usually employ a modular configuration described in by Dutta et al. [21]. The same reference provides a typical transfer function of the 40-channel demultiplexer with 3–6 dB IL and 100 GHz (0.8 nm) channel spacing.

In the visible spectrum, and thus within the application range of SI-POF, a vast variety of thin-film interference filters is available from various manufacturers. Even though not particularly intended for POF applications, the visible interference filters represent an attractive solution for POF demultiplexers, where wavelength selectivity, low IL, and high isolation are required.

A dichroic mirror is a special type of interference filter intended for the spatial separation or combination of light at different wavelengths. It is designed to operate at 45° AOI, such that a certain spectral range is transmitted, whereas the rejected wavelength range is reflected at 90° angle with respect to the incident optical axis. A commercial visible spectrum dichroic mirror has a transition slope between the transmission and reflection band of typically 30–40 nm (see **Figures 3** and **4**). This is significantly less steep compared to the standard interference filters designed for

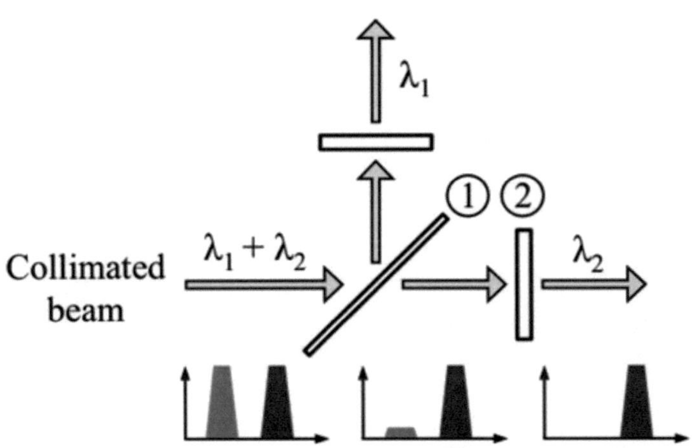

Figure 4.
Principle of separation of two collimated wavelength channels employing thin-film interference filters: (1) dichroic mirror (45° AOI); (2) interference filter (0° AOI).

the normal incidence. Unlike an interference filter, e.g., a long-pass mirror must be not only highly transmissive above the cutoff wavelength but also highly reflective below it. Therefore, producing steeper slopes would require increased complexity of the coating and, accordingly, a significant rise in production costs.

The interference filters show significant angular dependence of their transmission characteristic measured by Lee et al. [22]. To be applicable for SI-POF, the highly divergent beam from the fiber must be transformed into a bundle of parallel rays prior to the incidence. To increase the channel isolation, an additional band-pass filtering in each of the output channels should be implemented prior to the focusing of light. As an example, a selection of the dichroic mirror and interference filters for demultiplexing two wavelength channels centered around λ_1 = 450 nm and λ_2 = 525 nm is shown in **Figure 4**.

3.1 Data transmission techniques for increasing the channel capacity

Two data transmission techniques were used to overcome the bandwidth limitation of a POF WDM channel, which is primarily caused by the intermodal dispersion of SI-POF. Those were:

1. Non-return-to-zero (NRZ) modulation in combination with electronic dispersion compensation, in particular feed-forward equalization (FFE)

2. Discrete multitone (DMT) modulation

In the single-channel POF systems with intensity modulation and direct detection (IM/DD), those are well-known techniques for increasing the channel capacity. The next two subsections briefly introduce the two techniques: non-return-to-zero modulation and feed-forward equalization.

According to the Nyquist theorem for two-level signaling, the maximum bit rate (in bits per second) for a noiseless channel of the bandwidth B equals $2B$. Since the real systems encounter noise, this theoretical limit cannot be achieved. Therefore, for 100 m SI-POF link, the data rates less than 100 Mb/s are possible. If the data rate further increases, the eye diagram of the received signal becomes partially or completely closed due to the combination of the ISI and additive noise (introduced, e.g., by receiver's transimpedance amplifier).

The equalization techniques are used to open the eye diagram at the receiver for clock and data recovery (CDR). In its principle, the equalizer compensates for ISI, which is deterministic (unlike the random noise) and determined by the low-pass frequency response of a POF channel. In the work a simple linear FFE equalization technique was employed to correct the distorted signal waveforms at the receiver.

An FFE equalizer is realized as a discrete-time finite impulse response filter with adjustable coefficients. The output of the equalizer is obtained as the weighted sum of the delayed samples of the input signal as

$$V_e(t) = \sum_{k=0}^{N-1} c_k V_r(t - k T_D), \tag{6}$$

where $V_e(t)$ is the equalized voltage sample at the time t, c_k is the equalizer coefficient (weighting factor), T_D is the tap delay, and $V_r(t - k T_D)$ is the k times delayed uncorrected received voltage sample. A block diagram of the FFE equalizer is shown in **Figure 5**.

Through its coefficients the equalizer may synthesize a transfer function corresponding to the inverted channel frequency response, thus eliminating the ISI. In

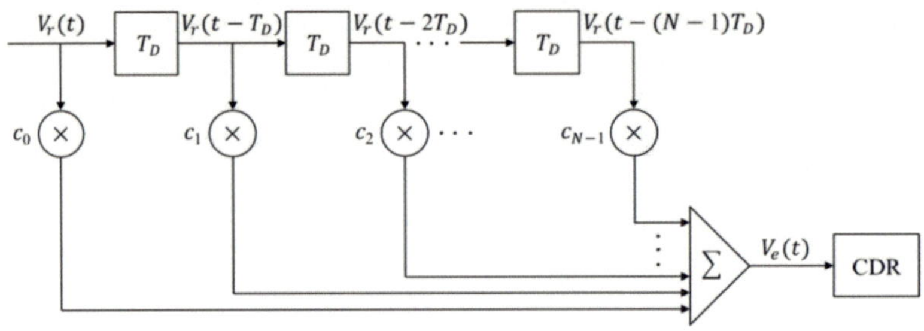

Figure 5.
Basic structure of an FFE equalizer.

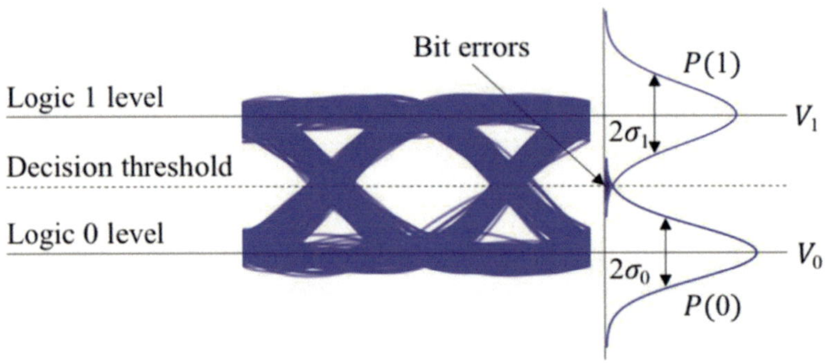

Figure 6.
Calculation of the Qy-factor from the eye diagram: P(1) *and* P(0), *probability distributions of received logic* 1 *and* 0 *levels, respectively.*

a noisy POF channel, and due to inverting the channel frequency response, this would lead to great noise amplification at higher frequencies where the channel frequency response is small in magnitude. Typically, to minimize the probability of the decision error, the weighting factors are calculated to minimize the noise power at the cost of a certain amount of residual ISI after equalization. The descriptions of different algorithms for optimizing the equalizer coefficients can be found in Loquai et al. [23].

The BER performance of the system was estimated based on the Q-factor of the equalized eye diagram (**Figure 6**)

$$Q = \frac{V_1 - V_0}{\sigma_1 - \sigma_0}, \qquad (7)$$

where V_1 and V_0 are the mean values and σ_1 and σ_0 are the standard deviations of the equalized signal voltages associated with logic 1 and 0 levels, respectively, as shown in **Figure 6**. The corresponding BER was obtained as

$$BER(Q) = \frac{1}{2} erfc\left(\frac{Q}{\sqrt{2}}\right), \qquad (8)$$

where $erfc(x) = (2/\sqrt{\pi}) \int_x^{\infty} exp(-\tau^2) d\tau$ is the complementary error function.

3.2 DMT: discrete multitone modulation

To provide high spectral efficiency of the signals transmitted within POF WDM channels, a DMT modulation technique was used. The DMT is a multicarrier modulation format and represents a baseband version of a better-known

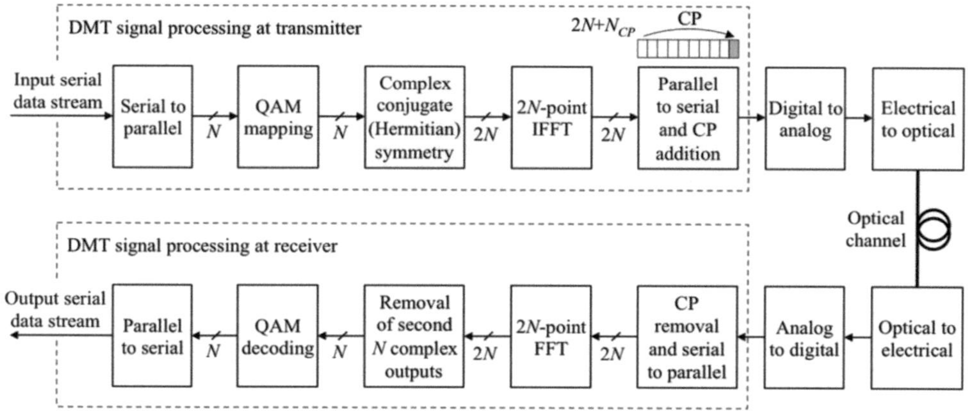

Figure 7.
Principle of DMT transmission over an optical IM/DD channel: N CP, length of CP.

orthogonal frequency division multiplexing (OFDM). Unlike the OFDM, which is used in wireless communication systems such as wireless local area networks (WLAN), the DMT is widely employed as enabling technology for digital subscriber lines (DSL), e.g., asymmetric DSL (ADSL) and very high DSL (VDSL). The DMT-based transmission was also shown to be very beneficial for SI-POF communication by Joncic et al. [24], Diaz et al. [25], and Vinogradov et al. [10].

The DMT technique slices the frequency-selective channel into a large number of subchannels that can be considered to have a flat frequency response. Each subchannel is then used for transmission of a passband signal with quadrature amplitude modulation (QAM). The simultaneous transmission of the low-speed parallel streams reduces the influence of the ISI. Another important property of DMT is that it adapts the signal parameters (QAM size and power in each subchannel) to the characteristic of the communication channel.

The principle of the DMT transmission over an optical IM/DD channel is shown in **Figure 7**. A high-speed serial data stream is first divided into N parallel lower-speed streams, where N corresponds to the number of subcarriers. Every M bits in each stream are grouped together and mapped into a complex value corresponding to a point of the QAM constellation with 2^M size. The modulation of the complex values onto N different subcarrier frequencies is based on the inverse fast Fourier transform (IFFT). To obtain real-valued time domain signal samples at the output of the IFFT block, a $2N$-point IFFE has to be carried out. Thereby, the second half of $2N$ IFFT inputs must be complex conjugate of the first half and symmetric around the middle of the input vector, a property referred to as a Hermitian symmetry. The real-valued samples of the DMT time signal are then parallel-to-serial converted, and a cyclic prefix (CP) is added as a guard interval at the beginning of each DMT frame to resist to inter-frame interference occurring in a band-limited channel. After digital-to-analog conversion, a bipolar continuous-time DMT frame is produced. It is then DC-biased to modulate the intensity of an electrical-to-optical converter.

At the receiver, the DMT waveform is direct-detected, analog-to-digital and serial-to-parallel converted, and demodulated using $2N$-point fast Fourier transform (FFT). The first N complex outputs of the FFT block are QAM-decoded and parallel-to-serial converted to result in an output serial data stream.

4. POF demultiplexer employing thin-film interference filters

This chapter focuses on the experimental realization of a thin-film interference filter-based SI-POF demultiplexer using a modular and precisely adjustable setup. In a step-by-step approach, the intermediate solutions with two and three channels were first established. In addition, two different configurations of the target demultiplexer setup with four channels were realized. The principle of operation and the approach for experimental realization are explained for the simplest case of a two-channel demultiplexer. The same basic principles also apply to the demultiplexers with higher channel count.

4.1 Operating principle and measurement results for a four-channel demultiplexer

By extending the channel count to four, it was possible to investigate two different demultiplexer configurations. Those were:

- Serial configuration
- Two-stage configuration

The principle of operation of a four-channel demultiplexer with serial configuration is shown in **Figure 8**. In this configuration the dichroic mirrors were cascaded such that each mirror (except the last one) demultiplexed a single-wavelength channel while passing all other wavelengths.

For practical realization 425, 505, and 567 nm cutoff long-pass dichroic mirrors were cascaded so that the interference filters centered at 405, 450, 525, and 650 nm could be implemented in the output ports 1–4, respectively. The corresponding transfer function and the basic parameters of the demultiplexer are shown in **Figure 9** and **Table 3**, respectively. The factors contributing to high IL in the output ports 3 and 4 are discussed in the next subchapter.

The principle of operation of a four-channel demultiplexer with two-stage configuration is shown in **Figure 10**. The first stage of the demultiplexer, represented by a dichroic mirror that follows directly after the collimating lens, splits the

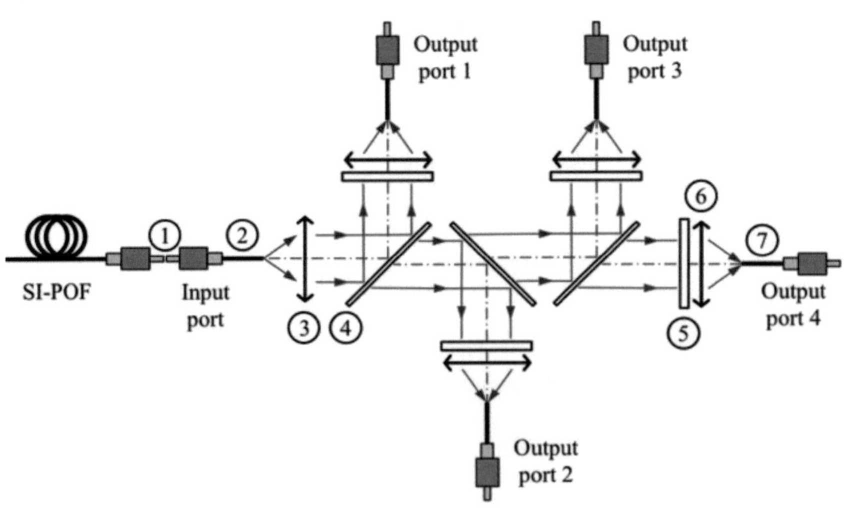

Figure 8.
Principle of operation of a four-channel SI-POF demultiplexer with serial configuration (see enumeration in **Figure 3***).*

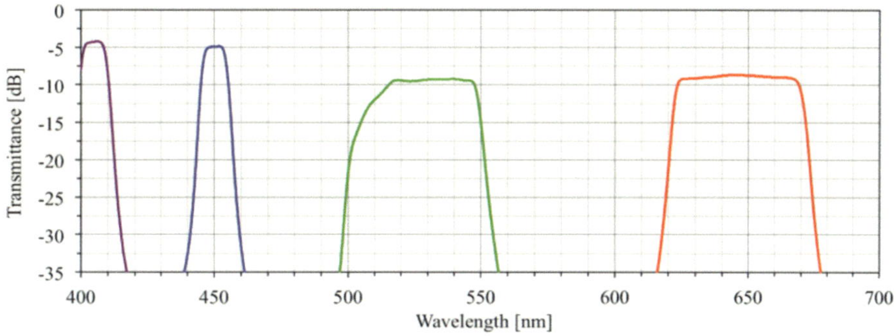

Figure 9.
Transfer function of the four-channel demultiplexer with serial configuration and the channels centered at 404.9, 450.1, 529.1, and 646.4 nm.

Output port	1	2	3	4
Center wavelength [nm]	404.9	450.1	529.1	646.4
3 dB passband bandwidth [nm]	9.4	9.2	39.6	47.8
Minimum IL [dB]	4.24	4.86	9.21	8.63
IL uniformity [dB]	4.97			
(Non)adjacent channel isolation [dB]	>30			

Table 3.
Basic parameters of the four-channel demultiplexer with serial configuration.

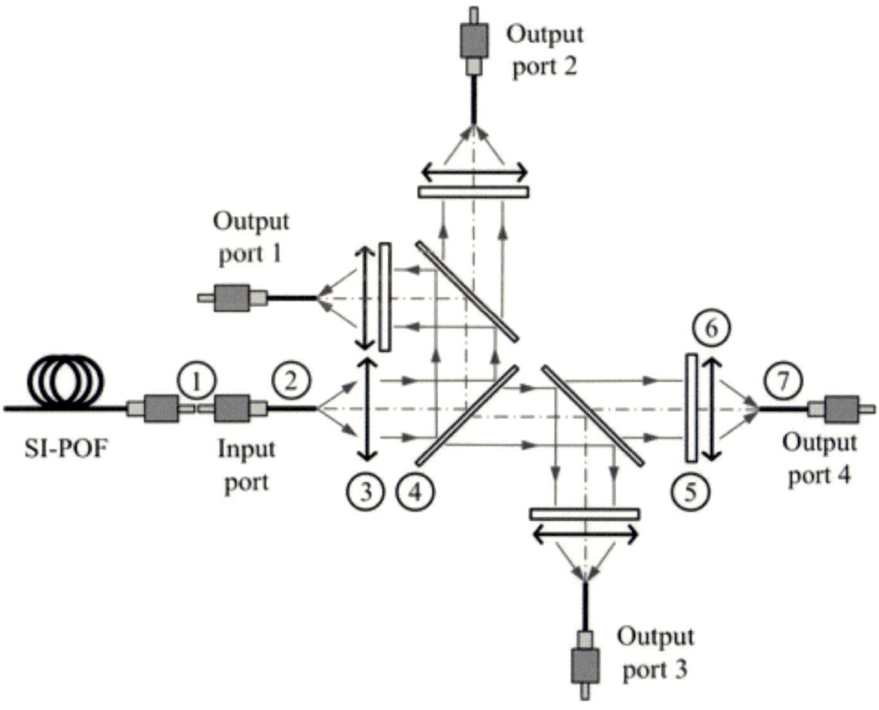

Figure 10.
Principle of operation of a four-channel SI-POF demultiplexer with two-stage configuration (see enumeration in **Figure 8***).*

incident spectrum into two spectral bands. The separation of the individual wavelength channels was then performed within the second stage of the demultiplexer. The corresponding laboratory setup is shown in **Figure 11**.

The practical realization was carried out with 505 nm cutoff long-pass dichroic mirror in the first stage of the demultiplexer. It reflected the lower spectral band so that 425 nm cutoff dichroic mirror was used in the second stage to demultiplex the signals for the output ports 1 and 2 in which 405 and 450 nm filters were employed, respectively. The upper spectral band transmitted by 505 nm mirror was demultiplexed in the second stage by 567 nm dichroic mirror. The filters centered at 525 and 650 nm were used in the output ports 3 and 4, respectively.

The corresponding transfer function is shown in **Figure 12**. The basic parameters of the demultiplexer are given in **Table 4**. The measurement results for the four-channel demultiplexer with two-stage configuration were presented at the International Conference on Plastic Optical Fibers (ICPOF) 2013 [26]. To comply with all other measurements shown in this chapter, which were performed 2 years thereafter, the demultiplexer setup was assembled and characterized

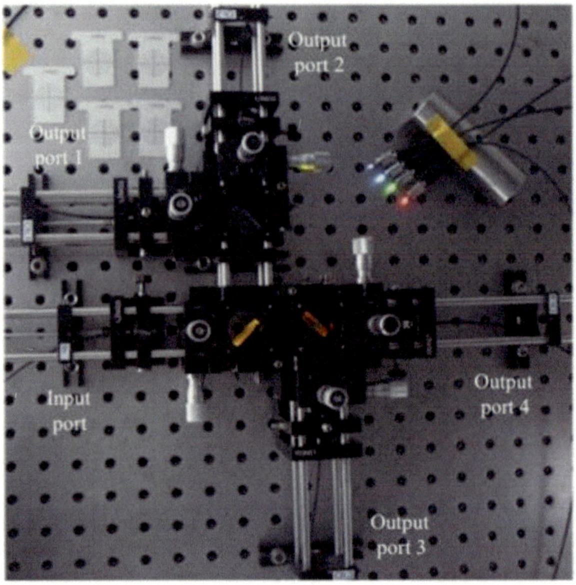

Figure 11.
Laboratory setup of the four-channel demultiplexer with two-stage configuration.

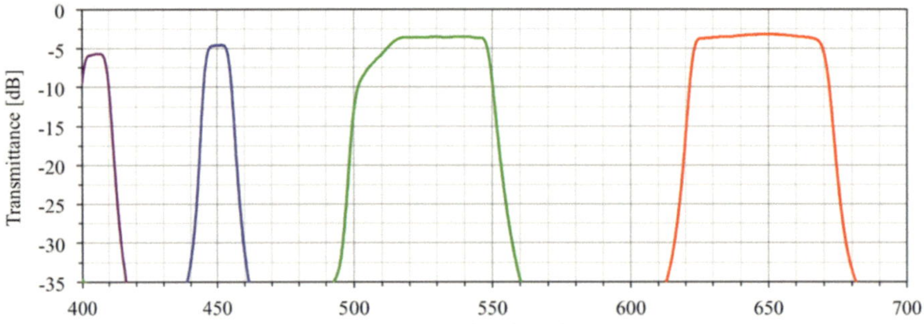

Figure 12.
Transfer function of the four-channel demultiplexer with two-stage configuration and the channels centered at 404.9, 450.1, 528.3, and 646.4 nm [26].

Output port	1	2	3	4
Center wavelength [nm]	404.9	450.1	528.3	646.4
3 dB passband bandwidth [nm]	9.4	9.3	41.6	47.6
Minimum IL [dB]	5.66	4.55	3.47	3.19
IL uniformity [dB]	2.47			
(Non)adjacent channel isolation [dB]	>30			

Table 4.
Basic parameters of the four-channel demultiplexer with two-stage configuration [26].

again. While preserving the same principal behavior of the spectral response, the minimum IL in the output ports 1 to 4 was 6.15, 5.44, 4.21, and 3.85 dB, respectively. Those were by 0.49, 0.89, 0.74, and 0.66 dB higher values than those reported by Appelt et al. [26].

5. Discussion

The transfer functions shown in **Figures 9** and **12** comprised for each demultiplexer channel:

- Loss of the connector interface at the demultiplexer input
- Attenuation of 1 m ingoing and 1 m outgoing SI-POF
- Propagation losses through the setup between the fiber end faces

The loss of the connector interface was minimized by applying the index-matching gel. This loss downscaled the transfer function of the demultiplexer by approx. 0.5 dB. To obtain the performance of the optomechanical setup itself, the value of 0.5 dB should be added to the measured transmittance values. The propagation losses included the Fresnel loss at the end face of the ingoing and the outgoing fiber, the losses introduced by the optical components (including reflections on the anti-reflection coatings), and the coupling losses due to the setup misalignments, optical aberrations, and clear aperture of components. For the perfectly aligned components and for given distances between them (obtained, e.g., from the CAD model), the minimum loss of the demultiplexer could be estimated by means of an optical ray tracing software. However, that work was beyond the scope of this work.

The shape of the spectral response of each demultiplexer channel was predominantly determined by an interference filter that was used. Those filters provided flat-top response, steep transition slopes, and high isolation between the channels due to an optical density greater than 4 (transmission of <0.01%) in the rejection bands within 400–700 nm region. The deviations of the channels from the nominal central wavelengths and bandwidths of interference filters comply with the center wavelength and passband bandwidth tolerances of ±2 and ±5 nm for 10 and 50 nm filters, respectively. An exception is the green channel where the spectral response curve was truncated by 505 nm cutoff dichroic mirror with the transmission band starting at 520 nm.

The four-channel demultiplexer introduced an additional channel in the short wavelength region. That allowed simultaneous operation at the violet and blue wavelengths, which are both very attractive for POF communication due to the availability of commercial laser diodes. Two different demultiplexer configurations offered significantly different performance.

In the serial configuration, the longer wavelength channels corresponded to the higher output ports. Because of the longer optical path than the shorter wavelength channels, the longer wavelength channels:

- Were more sensitive to alignment inaccuracies

- Encountered more optical components (dichroic mirrors)

- Suffered from stronger optical aberrations

- Experienced increased beam radius due to the beam divergence caused by the finite size of the source fiber [27]

The influence of those effects can be observed in the transfer function from **Figure 12**, where the green and red channels experienced significantly higher IL than the violet and blue ones. If the effect of alignment inaccuracy, which is a parameter related to the particular setup adjustment, would be disregarded, all other effects that are inherent to the serial configuration would lead to the same principal behavior of the transfer function.

The Appelt et al. [26] demultiplexer outperformed the four-channel solution from [28] in terms of IL and especially crosstalk. An exceptional performance of that demultiplexer with IL between 3.19 and 5.66 dB (overall minimum IL of 16.87 dB) may be explained by a very precise alignment of the components. However, all other measurements (performed 2 years thereafter) with two-, three-, and four-channel setups, which had to be each time newly aligned, showed somewhat higher IL but also very consistent behavior to one another. Therefore, it cannot be excluded that some other factors such as accumulated dust on the optical surfaces or coating damages due to improper handling could have introduced additional attenuation compared to [26] measurement, which was performed with brand new components. In spite of that, all subsequent measurement results, including the IL of 3.85–6.15 dB for the reassembled two-stage demultiplexer, can be considered as excellent achievements.

The significance of these and of the other previously realized interference filter-based SI-POF demultiplexers is that they enable realization of POF WDM systems and investigation on their data-carrying capacity. For these reasons it is important to further optimize the realized demultiplexer setup and extend the channel count.

6. WDM transmission employing NRZ modulation

This subchapter shows the initial experimental setup and gives the measurement results prior and after the first optimization step. Even though performed measurements cannot be considered as real WDM, the setup with multiplexer and demultiplexer along the optical path was assembled, and its functionality was demonstrated.

6.1 Experimental setup

The data transmission setup is shown in **Figure 13**. It comprised an Agilent N4903A bit error rate tester (BERT), four butt-coupled edge-emitting laser diodes, multiplexing POF coupler, 10 m SI-POF link, interference-based POF demultiplexer, optical receivers, and Agilent 86100B sampling oscilloscope.

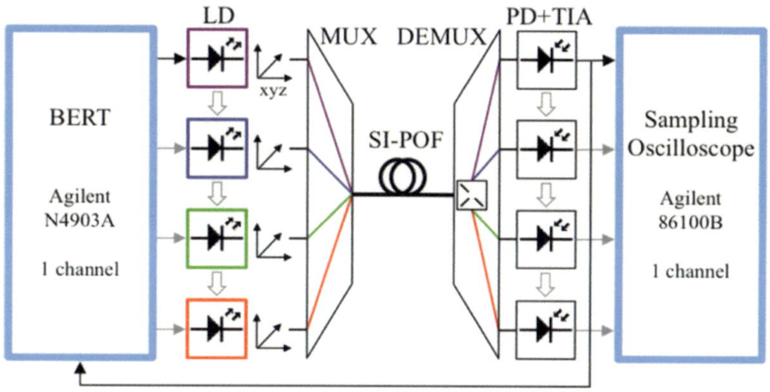

Figure 13.
Experimental setup for the measurements employing NRZ modulation: LD, laser diode; MUX, multiplexer; DEMUX, demultiplexer; PD, photodiode; TIA, transimpedance amplifier.

To provide precise temperature control, prevent possible damage from overheating, and extend the lifetime, the laser diodes were mounted in Thorlabs TCLDM temperature-controlled laser diode mounts. The temperature of an integrated temperature control (TEC) element of the mount was adjusted to +15°C. Only for OSRAM samples the cooling at +10°C was used to provide better stability of the optical output power. Both the temperature of the TEC element and the bias current were controlled over a Thorlabs ITC8022 module. Four of those modules were installed in a Thorlabs PRO8000 modular chassis for the simultaneous control of four operating diodes (see **Figure 15**). To maximize the coupling efficiency from the laser diode into the fiber, a butt-coupling unit based on an *xy*-translator (0.25 mm pitch of the adjustment screws) was utilized. The translator was mounted on the TCLDM9 mount over a 4-rod construction system, as shown in **Figure 15**. To combine the optical signals with different wavelengths onto the SI-POF link, a Comcore 4 × 1 fused POF coupler was used (see **Table 5**).

6.2 Measurement results and discussion

6.2.1 Gb/s transmission over 10 m SI-POF

The laser diodes providing the signals at four different wavelengths operated at 405 (DL-5146-101S), 450, 515, and 660 nm. Each diode was inserted into a laser socket of the TCLDM mount, which was also equipped with an internal 500 MHz bandwidth bias tee and had separate inputs for the bias and modulating current. The respective data rates achieved in the individual WDM channels were 0.5, 0.5, 0.7, and 0.8 Gb/s. The corresponding eye diagrams are represented in **Figure 14**. The effect of pulse shaping due to the low-pass characteristic of the fiber can clearly be recognized, e.g., in 405 nm channel.

Operating wavelength [nm]	405	450	515	639
Laser diode-to-coupler port launching loss [dB]	1.5	1.3	1.3	1.5
IL of 4 × 1 fused POF coupler [dB]	7.9	8.9	8.1	8.6
Connector loss (with index-matching gel) [dB]		0.5		
Total loss [dB]	9.9	10.7	9.9	10.6

Table 5.
Optical power loss at the transmitter side when using 4 × 1 fused POF coupler.

The irregular signal trajectories, e.g., in the eye diagrams of 515 and 660 nm channels, indicate the presence of nonlinearities in the electrical domain (presumably introduced by the bias tee circuit). No BER measurement data were saved. However, due to the eye diagrams still opened wide enough, it can be reasonably assumed that the corresponding BERs were below the FEC threshold of 10–3, allowing for the error correction.

High optical isolation of the demultiplexer provided very low optical crosstalk between the WDM channels. The crosstalk of ≤35 dB, coming from 450 nm channel, was detected in 515 nm channel. In all other channels, the crosstalk lower than −45 dB was detected with a Melles Griot 13 PDH 005 integrating sphere. Considering the amplitude levels of the recorded eye diagrams and low interchannel crosstalk, no reduction in the SNR of the received signals could be assumed if the laser diodes were modulated simultaneously. Therefore, it can be stated that an aggregate bit rate of 2.5 Gb/s could be transmitted over 10 m SI-POF with four

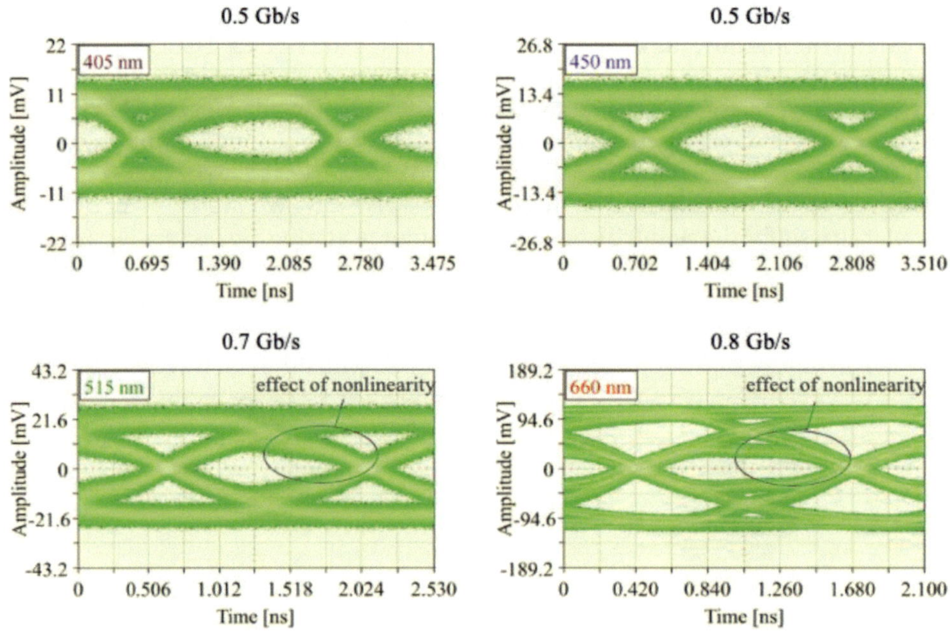

Figure 14.
Eye diagrams for 10 m SI-POF link at an aggregate bit rate of 2.5 Gb/s (note: The full time scale was automatically set by the oscilloscope and is smaller than two unit intervals of the signal).

(a) (b) (c)

Figure 15.
TO-56 diode mounted in a retainer ring; (c) TCLDM9 mount with a butt-coupling unit.

simultaneously active channels and no interchannel errors. The corresponding information rate after deduction of 7% FEC overhead would be 2.33 Gb/s.

6.3 5 Gb/s transmission over 10 m SI-POF

In the next experiment, 405 (DL-5146-101S), 450, 515, and 639 nm laser diodes were used as WDM optical sources. To directly modulate the diodes with a higher modulation bandwidth, each diode was soldered to a 50 ohm SMA formable coax cable (**Figure 14**), which was connected to the output of an external bias tee with 6 GHz bandwidth and 0.1 MHz low cutoff frequency.

Due to the low impedance of the laser diodes (typically 2–5 ohms), a severe impedance mismatch was present. However, by using higher power of the modulating signals, the mismatch could be compensated. For mounting the diodes into the mounts, copper retainer rings for both TO-56 and TO-38 packages were fabricated (**Figure 15**). For better thermal conductivity between the TEC element of the mount, the retainer ring, and the laser diode housing, a heat-conductive paste was applied on the contact surfaces.

The bit rates in the individual channels were 1.25 Gb/s (405 nm channel), 1.05 Gb/s (450 nm channel), 1.25 Gb/s (515 nm channel), and 1.45 Gb/s (639 nm channel). The transmission parameters for the individual channels are listed in **Table 6**. In contrast to the previous measurement, no signal nonlinearities were present, and the maximum achievable data rates were limited by the ISI.

Taking into account the amplitude levels of the recorded eye diagrams and low interchannel crosstalk, like in the previous experiment, no power penalty due to crosstalk could be assumed if the laser diodes were modulated simultaneously. Therefore, it can be stated that 5 Gb/s transmission could be realized over 10 m SI-POF link at the BER $<10^{-4}$ with four simultaneously active channels and no interchannel errors. Using a standard Reed-Solomon (255,247) FEC with 3.2% redundancy, 4.84 Gb/s transmission could be achieved at the BER$<10^{-9}$ [28].

6.4 WDM transmission employing NRZ modulation and FFE equalization

This subchapter shows the improved experimental setup and gives the measurement results of the simultaneous four-channel NRZ transmission over 50 m SI-POF. To mitigate the effects of ISI, the FFE equalization was implemented at the receiver side. The experimental setup is shown in **Figure 16**. It comprised a four-channel Agilent M8190A arbitrary waveform generator (AWG), four butt-coupled edge-emitting laser diodes, four-legged multiplexing POF bundle, SI-POF link of two different lengths, interference-based POF demultiplexer, Graviton SPD-2 receiver, and four-channel Agilent DSA91604A real-time oscilloscope with built-in software for digital signal processing. A photo of the general setup for investigating four-channel high-speed POF WDM transmission is shown in **Figure 17**.

To multiplex the signals from four laser diodes onto the SI-POF link, a four-legged POF bundle was used. A multiplexing interface is formed by positioning the fiber bundle against 1 mm SI-POF.

Operating wavelength [nm]	405	450	515	639
Bit rate [Gb/s]	1.25	1.05	1.25	1.45
BER	$6{\cdot}10^{-5}$	$7{\cdot}10^{-5}$	$1.1{\cdot}10^{-6}$	$9{\cdot}10^{-7}$

Table 6.
Transmission parameters for 10 m SI-POF link at an aggregate bit rate of 5 Gb/s.

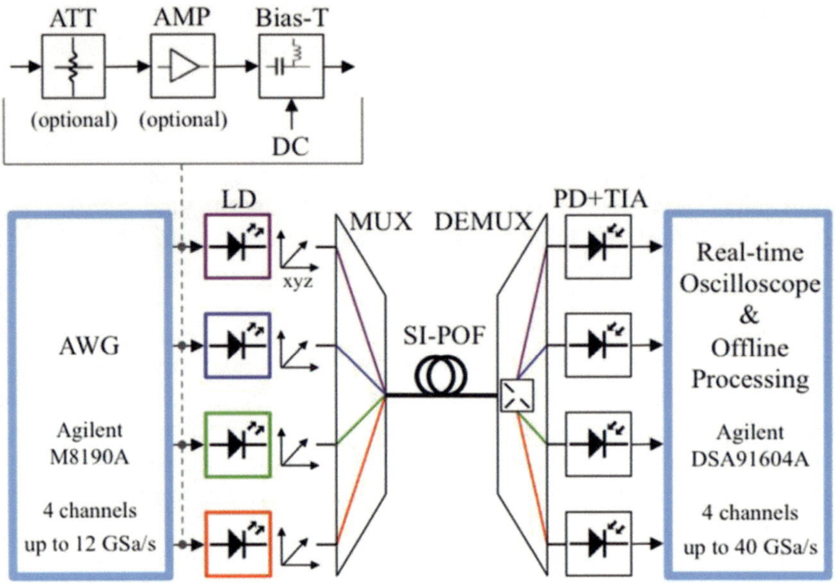

Figure 16.
Experimental setup for the measurements employing NRZ modulation and offline-processed FFE: ATT, attenuator; AMP, amplifier; DC, direct current.

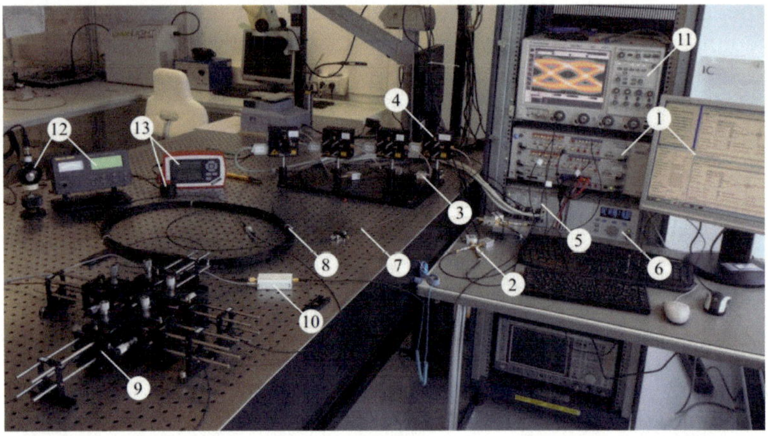

Figure 17.
Experimental setup for investigating four-channel high-speed POF WDM transmission: (1) Agilent M8190A AWG; (2) attenuator and MERA-556+ wideband amplifier; (3) bias tee; (4) Thorlabs TCLDM temperature-controlled laser diode mount, (5) Thorlabs ITC8022 module; (6) Thorlabs PRO8000 modular chassis; (7) four-legged multiplexing POF bundle; (8) SI-POF link; (9) four-channel interference filter-based demultiplexer; (10) graviton SPD-2 receiver; (11) Agilent DSA91604A real-time oscilloscope; (12) Melles Griot universal optical power meter with 13 PDH 005 integrating sphere; (13) Thorlabs PM100D power meter with S140C integrating sphere.

For realization of the bundle, an Asahi KASEI DB-400 PMMA SI-POF with 400 μm cladding diameter and NA = 0.5 was used. Four 60-cm-long fibers were terminated at the input side with 400 μm FC connectors. The opposite ends of the fibers were joined together and glued inside 970 μm FC connector to form the fiber bundle (**Figure 18**). As illustrated in **Figure 18**, an FC connector-mating sleeve was used to bring together and align the bundle and the input of the SI-POF link, thereby forming the multiplexing interface. An index-matching gel was applied between the connectors to reduce the losses.

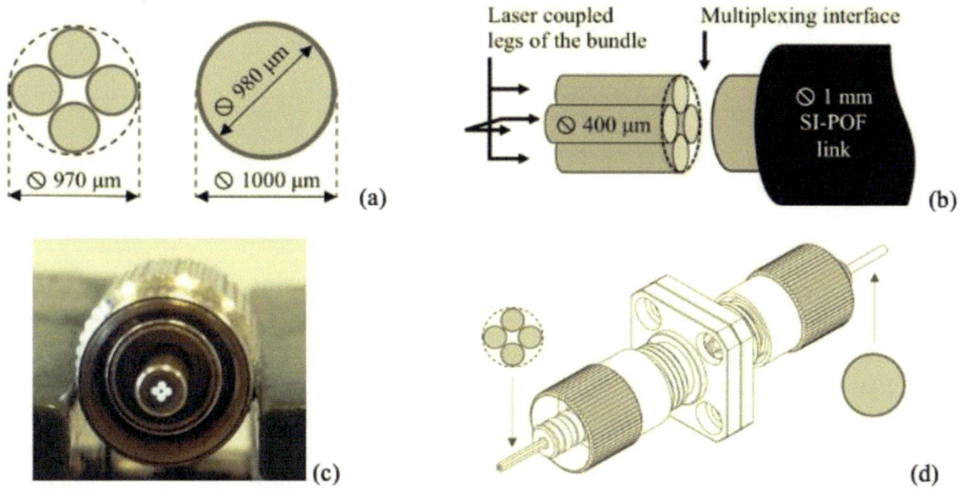

Figure 18.
Four-legged multiplexing POF bundle: (a) cross sections of four 400 µm cladding diameter fibers arranged within a circle with 970 µm diameter (left) and of 980/1000 µm SI-POF (right); (b) principle of operation of the POF bundle as a multiplexer; (c) four 400 µm fibers glued within 970 µm FC connector; (d) formation of a multiplexing interface with the POF bundle aligned against the SI-POF link using an FC connector-mating sleeve.

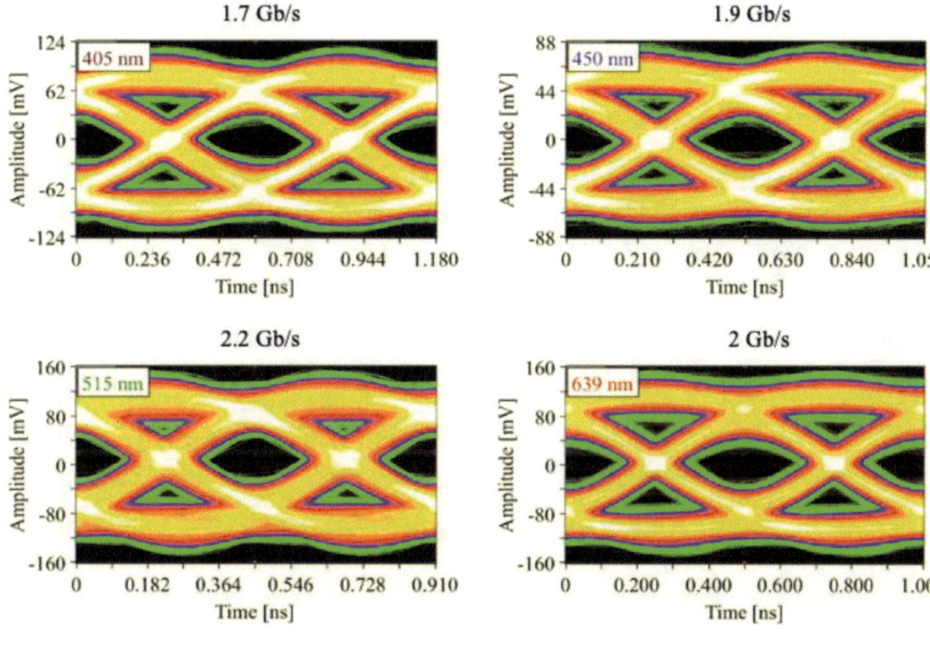

Figure 19.
Eye diagrams for 50 m SI-POF link at an aggregate bit rate of 7.8 Gb/s.

The described multiplexing solution was first shown in [29]. Shortly before, the patent application for an optical POF multiplexer based on a multi-legged POF bundle, which referred to arbitrary channel counts and fiber diameters, was submitted to the German Patent and Trade Mark Office (DPMA) under number DE 102013 020236.1. A similar approach was later adopted in [30, 31] to realize the low loss seven-legged and three-legged multiplexers, respectively.

The AWG simultaneously generated four independent NRZ data streams (**Figures 16** and **19**) based on 27-1 PRBS with the maximum sampling rate.

Operating wavelength [nm]	405	450	520	639
Laser diode-to-400 μm fiber launching loss + attenuation of a bundle leg [dB]	4.06	2.63	2.82	4.76
Connector loss (with index-matching gel) [dB]	0.87	0.98	0.52	0.77
Total loss [dB]	4.93	3.61	3.34	5.53

Table 7.
Optical power loss at the transmitter side when using four-legged POF bundle.

The received electrical signals were acquired by the real-time oscilloscope with 8-bit vertical resolution and oversampling. The digital receiver equalization was carried out in the offline mode. For that purpose the oscilloscope's built-in Serial Data Equalization software was used [32].

To prevent the equalizer from amplifying the noise components at higher frequencies where the energy content of useful signal was low, the bandwidth of the oscilloscope was set to the value in GHz corresponding to one half of the data rate in Gb/s. A phase-locked loop was used to extract the clock from the equalized data.

The eye diagram of an equalized waveform was displayed on the oscilloscope's screen for further analysis. The oscilloscope's built-in software EZJIT Complete was used to estimate the corresponding Q-factor [33]. Thereby, only a small time window (2% of the unit interval) in the middle of the equalized eye diagram was taken into consideration. The BER value was then calculated using Eq. (5). **Table 7** shows the optical power losses of the four used WDM channels.

6.5 7.8 Gb/s transmission over 50 m SI-POF

The WDM channels based on 405 nm (DL-7146-101S), 450, 515, and 639 nm laser diodes were employed in this transmission experiment. To maximize the modulating signal amplitude and thereby improve the SNR of the received signal, the optical output power of each laser diode was adjusted to its maximal possible value. For 405, 450, and 515 nm devices, the operating point was set to comply with the upper limit of the receiver's dynamic range. The 639 nm device was driven with the maximum recommended forward current. The respective used bias currents were 70, 40, 61, and 43 mA. The optical powers coupled into the SI-POF link and the received optical powers measured after the demultiplexer are given in **Table 7**. All diodes were driven in their linear lasing region. The 1 Vpp output amplitude of the AWG was sufficient to modulate 450, 515, and 639 nm laser diodes with the modulation index of approx. 0.9. The signal in 405 nm channel was additionally amplified to achieve the same modulation index. For the amplification, a MERA-556+ wideband amplifier (20.5 dB gain at 0.1 GHz) was used in combination with 10 dB attenuator to avoid amplifier nonlinearities (18 dBm output power at 1 dB compression point at 0.1 GHz).

The maximum transmission rates achieved in the individual channels were 1.7 Gb/s (405 nm channel), 1.9 Gb/s (450 nm channel), 2.2 Gb/s (515 nm channel), and 2 Gb/s (639 nm channel). Thereby, six FFE taps with the tap delay equal to one half of the corresponding bit period were used in each of the channels. The transmission parameters for the individual channels are listed in **Table 8**. The resulting eye diagrams are represented in **Figure 19**. A total of 7.8 Gb/s were transmitted over 50 m SI-POF at the BER < 10^{-5}. After deduction of 3.2% redundant bits required for Reed-Solomon (255,247) FEC, a net bit rate of 7.56 Gb/s was obtained. Compared to the record capacity of a single-channel system over the same fiber length used by Vinogradov et al. [10], an improvement of 1.67 Gb/s was achieved.

Operating wavelength [nm]	405	450	515	639
Bit rate [Gb/s]	1.7	1.9	2.2	2
Q-factor	5.06	4.67	4.96	4.76
BER	$2.10{\cdot}10^{-7}$	$1.51{\cdot}10^{-6}$	$3.52{\cdot}10^{-7}$	$9.68{\cdot}10^{-7}$

Table 8.
Transmission parameters for 50 m SI-POF link at an aggregate bit rate of 7.8 Gb/s.

7. Conclusions

We experimentally demonstrated the feasibility and potential of a high-speed POF WDM concept; a four-channel data transmission setup was realized. A four-legged multiplexing POF bundle was developed to combine the signals from four visible laser diodes onto SI-POF link. For the separation of wavelength channels, the interference filter-based demultiplexer with two-stage configuration was used. It was shown that POF WDM with lower channel rates and simple transmission technique (NRZ + FFE) could provide aggregate bit rates comparable to those achieved with the single-wavelength systems that used advanced modulation formats (DMT or PAM + DFE) and required significant signal processing. In addition, the 50 m SI-POF link at an aggregate bit rate of 7.8 Gb/s was demonstrated over 50 m SI-POF, respectively, at the BER = 10^{-3}.

Acknowledgements

We gratefully acknowledge the funding by the German Ministry of Education and Research (BMBF) under grant number 16V0009 (HS Harz)/16V0010 (TU BS). All injection molded parts are done with the support of the Institute of Micro and Sensor Systems at the Otto von Guericke University Magdeburg and Prof. Bertram Schmidt.

Author details

Ulrich H.P. Fischer-Hirchert* and Mladen Joncic
Harz University of Applied Sciences, Wernigerode, Germany

*Address all correspondence to: ufischerhirchert@hs-harz.de

References

[1] Fischer-Hirchert UHP, Haupt M, Joncic M. Optical transmission systems using polymeric fibers. In: Predeep P, editor. Optoelectronics—Devices and Applications. Intech Open; 2011 Chapter 22

[2] IEC09, Specification T. TS 105 175-1—V1.1.1—Access, Terminals, Transmission and Multiplexing (ATTM); Plastic Optical Fibre System Specifications for 100 Mbit/s and 1 Gbit/s. 2010. 1: pp. 1-16

[3] Grzemba A. MOST. The Automotive Multimedia Network. From MOST25 to MOST150. Poing: Franzis; 2011. p. 250

[4] Access, Terminals, Transmission and Multiplexing (ATTM); Plastic Optical Fibres; Part 1: Plastic Optical Fibre System Specifications for 100 Mbit/s and 1 Gbit/s; Sub-Part 2: 1 Gbit/s and 100 Mbit/s Physical Layer for Plastic Optical Fibres, ETSI Technical. 2015

[5] Unified high-speed wireline-based home networking transceivers—System architecture and physical layer specification, ITU-T Recommendation G.9960. December 2011

[6] Vinogradov J, Ziemann O, Lednicky O, Gottschalk J, Zceh M, Tchoupkoua S. Optimal equalizers for SI-POF and Gpbs. In: 17th Int Conf Plast Opt Fibers. 2008

[7] Loquai S, Kruglov R, Schmauss B, Bunge C-A, Winkler F, Ziemann O, et al. Comparison of modulation schemes for 10.7 Gb/s transmission over large-core 1 mm PMMA polymer optical fiber. IEEE Journal Light Technology;**31**(13):2170-2176

[8] Breyer F. Multilevel transmission and equalization for polymer optical fiber systems. Ph.D. dissertation, Germany: Fakultät für Elektrotechnik und Informationstechnik, Technische Universität München; 2010

[9] Lee SCJ. Discrete multitone modulation for short-range optical communication. Ph.D. dissertation, Netherlands: Department of Electrical Engineering, Eindhoven University of Technology; 2009

[10] Vinogradov J, Kruglov R, Loquai S, Ziemann O. Multi gigabit transmission with blue, green and red laser diodes. In: 20th International Conference on Plastic Optical Fibers; 2011. pp. 467-470

[11] Joncic M, Haupt M, Fischer UHP. Investigation on spectral grids for VIS WDM applications over SI- POF Untersuchung von Spektralen Gittern für VIS WDM-Anwendungen über SI-POF. In: ITG Fachtagung Photonische Netze. Leipzig: VDE Verlag; 2014. p. 38855

[12] Lekishvili N, Nadareishvili L, Zaikov G, Khananshvili L. Polymers and Polymeric Materials for Fiber and Gradient Optics. Amsterdam: Brill Academic Publishers; 2002

[13] Bunge C-A. Private communication—spectral attenuation of Asahi fiber specified for MOST

[14] Koike Y. Fundamentals of Plastic Optical Fibers. Hoboken, New Jersey: Wiley and Sons, Inc.; 2014

[15] Kaino T. In: Jones W, editor. Organic Molecular Solids: Properties and Applications. London: CRC Press; 1997. p. 352

[16] Emslie C. Review polymer optical fibers. Journal of Materials Science. 1988;**23**:2281-2293

[17] Groh W. Overtone absorption in macromolecules for polymer optical fibers. Die Makromol Chemie. 1988;**189**(12):2861-2874

[18] Ziemann O, Krauser J, Peter EZ, Daum W. POF Handbook—Optical Short Range Transmission Systems. Berlin Heidelberg: Springer Verlag; 2008. p. 455

[19] Thiele H-J, Nebeling M. Coarse Wavelength Division Multiplexing: Technologies and Applications. London: CRC Press, Taylor and Francis Group; 2007

[20] LIGHTEL Product specification. 4, 8, and 16 Channel Extended Band CWDM Mux/Demux. 2014

[21] Dutta AK, Dutta NK, Fujiwara M. WDM Technologies: Passive Optical Components. Cambridge: Acadamic Press; 2003

[22] Herbert V. Wavelength Filters in Fibre Optics. Berlin, Heidelberg: Springer; 2006

[23] S. Proakis. Digital Communications. 5th ed., McGraw-Hill; 2008.

[24] Lee SCJ, Breyer F, Randel S, Gaudino R, Bosco G, Bluschke A, et al. Discrete multitone modulation for maximizing transmission rate in step-index plastic optical fibers. IEEE Journal Light Technology. 2009;**27**(11):1503-1513

[25] Loquai S, Kruglov R, Ziemann O, Vinogradov J, Bunge C.-A. 10 Gbit/s over 25 m plastic optical fiber as a way for extremely low-cost optical interconnection. In: OFC/NFOEC. 2010. paper OWA6

[26] Joncic M, Höll S, Haupt M, Caspary R, Fischer UHP. Development status of a four-channel CWDM system for multi-Gbit/s data links over SI-POF. In: 22nd International Conference on Plastic Optical Fibers; 2013. pp. 59-64

[27] Diaz E, Knobl M. Prototyping illumination systems with stock optical components. Photonik International. 2012;**2**:4-7

[28] Appelt V, Vinogradov J, Ziemann O. Simple FEXT compensation in LED based POF-WDM systems. In: 11th International Conference on Plastic Optical Fibers; 2002. pp. 127-129

[29] Jončić M, Haupt M, Fischer UHP. Development status of a four-channel CWDM system for Multi-Gbit/s in-house data communication via SI-POF. In: 20 ITG-Fachtagung Kommun. 2013; pp. 6

[30] Kruglov R, Vinogradov J, Loquai S, Ziemann O, Bunge C-A, Hager T, Strauss U. 21.4 Gb/s discrete multitone transmission over 50-m SI-POF employing 6-channel WDM. In: OFC/NFOEC; 2014. p. Paper Th2A.2

[31] Pinzon PJ, Vaszquez C, Perez I, Lallana PC. Design and analysis of a WDM system for multi-Gbit/s transmission over 50 m of SI-POF. In: 23rd International Conference on Plastic Optical Fibers; 2014

[32] Agilent Technologies I. User's guide. Agilent N5461A Infiniium Serial Data Equalization [Online]. 2009

[33] Agilent Technologies I. Data sheet. EZJIT Complete Jitter and Vertical Noise Analysis Software for Infiniium Oscilloscopes [Online]. 2013

Section 3

Optical Processing and Characterization

Chapter 4

All Optical Signal Processing Technologies in Optical Fiber Communication

Muhammad Irfan Anis

Abstract

Due to continued growth of internet at starling rate and the introduction of new broadband services, such as cloud computing, IPTV and high-definition media streaming, there is a requirement for flexible bandwidth infrastructure that supports mobility of data at peta-scale. Elastic networking based on gridless spectrum technology is evolving as a favorable solution for the flexible optical networking supportive next generation traffic requirements. Recently, research is centered on a more elastic spectrum provision methodology than the traditional ITU-T grid. The main issue is the requirement for a transmission connect, capable of accommodating and handling a variety of signals with distinct modulation format, baud rate and spectral occupancy. Segmented use of the spectrum could lead to the shortage of availableness of sufficiently extensive spectrum spaces for high bitrate channels, resulting in wavelength contention. On-demand space assignment creates not only deviation from the ideal course but also have spectrum fragmentation, which reduces spectrum resource utilization. This chapter reviewed the recent research development of feasible solutions for the efficient transport of heterogeneous traffic by enhancing the flexibility of the optical layer for performing allocation of network resources as well as implementation of optical node by all optical signal processing in optical fiber communication.

Keywords: defragmentation, format conversion, grooming, multicasting, optical signal processing, semiconductor optical amplifier, wavelength conversion

1. Introduction

In recent years, with the proliferation of cloud computing and high-definition media streaming increasing the use of communications and information technology in photonic infrastructure. Fiber capacity crunch concerns are driving optical networking toward a spectral-efficiency-conscious design philosophy. Moreover, as the number of various high-capacity services increases such as video delivery service and data centers, transport network flexibility will become important, resulting in the demand of elastic transport optical networking [1–5]. Key methods for empowering the elastic/flexible optical system are: single SDN XCVR, BW increase, flexible spectrum approach, Multicore fiber using the SDM technology. Nonetheless, it would be cost-restrictive to convey an entirety set of advances.

Therefore, in the next generation optical communication, optical nodes will need to allocate resources in an elastic and effective way to professionally provision high and low data rate signals [6]. To represent this evolving network scenario, an elastic network is proposed and demonstrated, providing elastic resource allocation in spectrum as a means to address the disparity between required and allocated bandwidth. By using this technique it is possible to assign a customized bandwidth per channel depending on specific requirements. Elastic allocation in the spectral domain implies that the standard 50-GHz ITU grid is not used and a continuous spectrum can be allocated to accommodate high-capacity channels with large bandwidth requirements. Also, channels that require lower bandwidths can be accommodated more efficiently by using narrower channel spacing as long as the performance is not compromised. Next generation optical networking is expected to cope with a number of challenges, especially in terms of elastic optical nodes [7], as there is always a need of increasing flexibility in the allocation of spectral and temporal resources so that they are able to efficiently support on-demand services and functionality.

Apart from the obvious advances in capacity and performance, it is clear that with each progressive stage of evolution of the optical node, additional flexibility has been introduced, e.g. the introduction of wavelength granularity, the ability to add/drop individual wavelengths, reconfigurability, etc. Other types of functionality may also add flexibility to the system, e.g. wavelength conversion (WC), format conversion (FC), multicasting (MC), regeneration, etc. In [8] a broadcast and select node based on bandwidth-variable wavelength selective switches (BV-WSS) was proposed. However, the very nature of this architecture will restrict upgradeability and will limit support for evolving requirements and new functionalities, e.g. optical signal processing. However, in Reconfigurable Optical Add Drop Multiplexer (ROADM) architectures [9] it is difficult to introduce additional functionality due to the fact that several wavelengths are simultaneously switched over the same port. On the contrary, OXCs support additional functionality more naturally as wavelengths are split and switched individually. Thus, modules with the required functionality to operate on individual wavelengths can be positioned in the right place within the OXC. However, the requirement for a particular signal processing function is often uncertain, e.g. it may be required for some wavelengths at some time period and for other wavelengths at a different time period. Therefore, modules that provide a per-channel functionality are generally deployed for all wavelengths as there is no possibility of sharing them among several optical paths inside the OXC. A better solution would enable modules to be shared, thus improving modules' utilization and reducing the amount of modules required to satisfy a given demand for the offered functionality, i.e. better hardware utilization and efficiency.

WDM networks utilize routing and wavelength allocation (RWA) algorithms to find available resources for new requests. However, in elastic optical networks the problem is more difficult due to the new flexible spectrum allocation, where elastic spectrum bands rather than single wavelengths are considered. For new requests with specific bandwidth requirements, routing and spectrum allocation (RSA) algorithms need to identify sufficiently wide spectrum slots that are available from source to destination. Furthermore, as channels are added and removed, they leave behind noncontiguous slots of free spectrum. Although these fragments may add up to a considerable amount of bandwidth, new channel requests may be blocked due to the lack of sufficient contiguous spectrum [10]. Spectrum fragmentation may be prevented to some extent by introducing appropriate policies in RSA algorithms. Alternatively, techniques to defragment the spectrum may also be utilized, e.g. using wavelength conversion.

Another issue of optical node is that, it should manage transport of blend of suppliers' traffic facilities for legacy signals, core traffic and multiple format signals with variable bit rate. Thus, flexible optical nodes will ought to assign resources in an adaptable and proficient way to back a blend of super-channels and lower speed channels. The nodes' density will mainly influenced on the link interoperability between segments (i.e. core segment: need signal processing and to support super-channels, metro segments: might carry legacy 10 Gb/s [11] links).

To address increasing traffic growth, the most straightforward and economical way in which this can be done is to deploy additional 10G wavelengths. Thus, new 10G wavelengths are placed 50 GHz, according to the standard WDM grid, until the available bandwidth is exhausted. However, the maximum capacity that can thus be provided is 800 Gb/s (i.e. 80 × 10 G using only the C band), which is already insufficient for heavily used backbone network links. Furthermore, providing additional capacity in this manner is highly inefficient in terms of the spectral resources that are consumed. The immediate solution to this problem is to deploy 100 G links, despite their higher cost compared to 10 × 10 G. 100 G is more spectrally efficient than 10G as it can fit in a standard 50-GHz WDM slot, advance modulation formats (like DP-QPSK), coherent detection, extensive use of Forward Error Correction (FEC) and electronic impairment mitigation. However, this solution is expected to be viable only for the short and medium terms. For super-channels at 400 Gb/s [12], 1 Tb/s [13] and beyond [14, 15] will occupy broader spectrum, require more complex multilevel modulation formats with higher OSNR and consequent shorter reach which neither fits within the existing ITU grid nor is supported by conventional optical network infrastructures. For instance, optical cross-connects and ROADMs allocate only discrete 50-GHz slots of bandwidth due to their internal WDM (de)multiplexers. Channels that require wider bandwidths are severely distorted if passed through such devices. Therefore, in order to efficiently support high-speed channels, a flexible bandwidth infrastructure is required.

In spite of the increasing popularity of elastic optical networks, there has been very little work focusing on elastic node architectures. Therefore it is critical to investigate the details of realizing optical networking solutions toward flexible and efficient allocation of network resources. Further, provide fully functional intelligent infrastructure for simultaneously supporting the switching and transport of combination of high-capacity super-channels and lower bit rate channels [16]. How elastic nodes support dynamic and on-demand provisioning of functionality, such as spectrum defragmentation, wavelength conversion, regeneration, grooming, format conversion, time multiplexing, etc. leaves the door open for future research.

2. The roadmap toward elastic optical networks

A number of different industry surveys indicate the global IP traffic is increasing at a startling rate i.e. more than 30% per year [17] and estimates that by 2025 the Internet will burn 7% of the 2010 global electricity supply [18]. This heralds the start of a huge wave of data driven mainly by broadcasting or multicasting, streaming of IPTV, high-quality videos using ultrahigh-definition (4 k _ 8 k pixels) videos and rich media files that clients migrating to an all smart phones and tablets, enabling video to be consumed more conveniently stored in cloud architecture such as Microsoft Azure or Apple's iCloud via network connections anywhere, anytime.

To cope with this expected growth in traffic volume technological advances to date have allowed an increase in DWDM data rates to higher than 100 Gb/s per optical carrier, but these technologies will soon be close their practical or theoretical limits. To offset this growing trend and the consequences of unbridled demand, research efforts are focused on the ways to improve the efficiency of these networks, often by leveraging photonic alternatives to provide improved performance [19]. The evolution of optical nodes and networks has been characterized by continuous enhancements in the parameters listed in **Figure 1**. These parameters are inter-dependent and their relative temporal evolution, combined with the related network economics, will dictate the exact network evolution, assuming maturity of the available technology. For an optical network to accommodate all the above requirements, it must transparently transmit, switch channels (e.g. wavelengths,

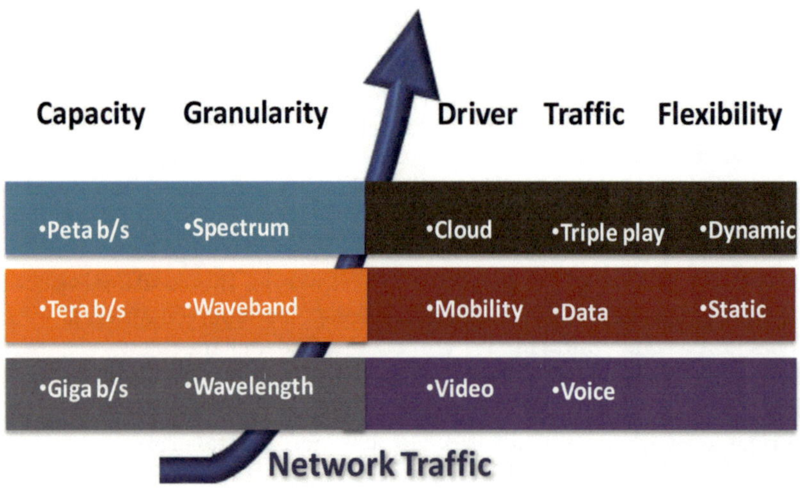

Figure 1.
Vision for a transparent reconfigurable optical network: it is the relative development of the each of five parameters with respect to time and that will dictate the network evolution.

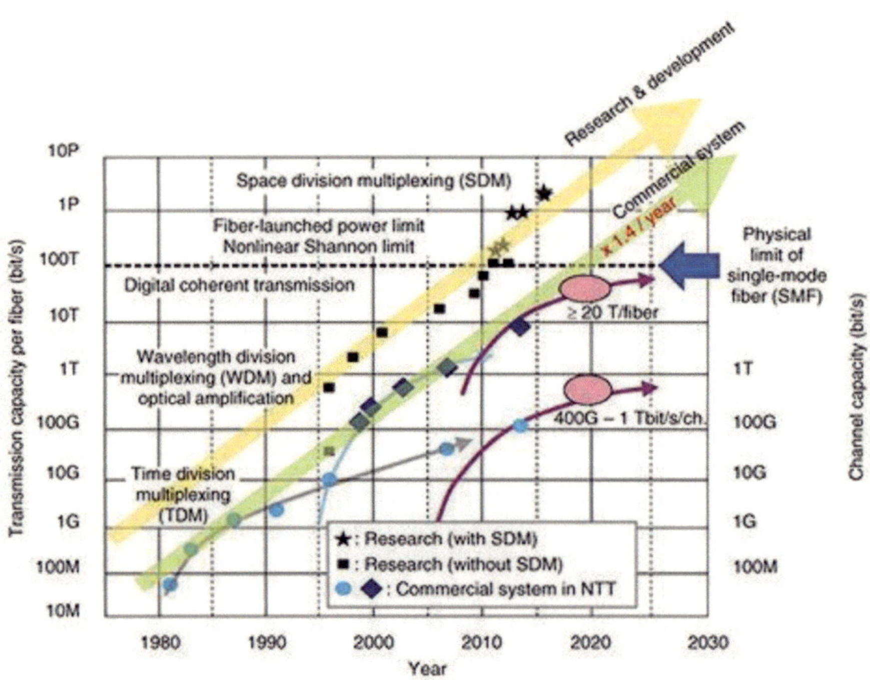

Figure 2.
Evolution in optical transmission technology [22].

wavebands, sub wavelength channels) and provide on-demand bandwidth in a scalable and reconfigurable fashion [20].

The graph (**Figure 2**) shows that the evolution of transmission capacity x fiber link has been growing over the years. The optical fiber bandwidth utilization approaches its peak limit quickly. Given the potential for such capacity crunch, the research community has concentrated on finding alternatives that make the most of the scarce network resources and meet the consistently expanding traffic requests [21]. In such context, adaptability or reconfigurable capability of networks will become more and more critical and hence spectrum efficient optical networking techniques have been introduced as a way to offer efficient utilization of the available optical resources. The place to start is with the transport network, which forms the foundation of elastic networking.

Currently, all deployed optical transport technologies are mainly based on a fixed grid 50 GHz or 100 GHz/frequency grid standardized by ITU-T and the same modulation technology has been used for optical signals at the same bitrate regardless of the transmission distance. In this scenario, the system is reaches its limits in terms of both capacity and flexibility [10, 23, 24] as higher capacities per fiber have been achieved by improving the spectral efficiency (SE) through increasing the bitrate per channel while keeping or even narrowing the channel

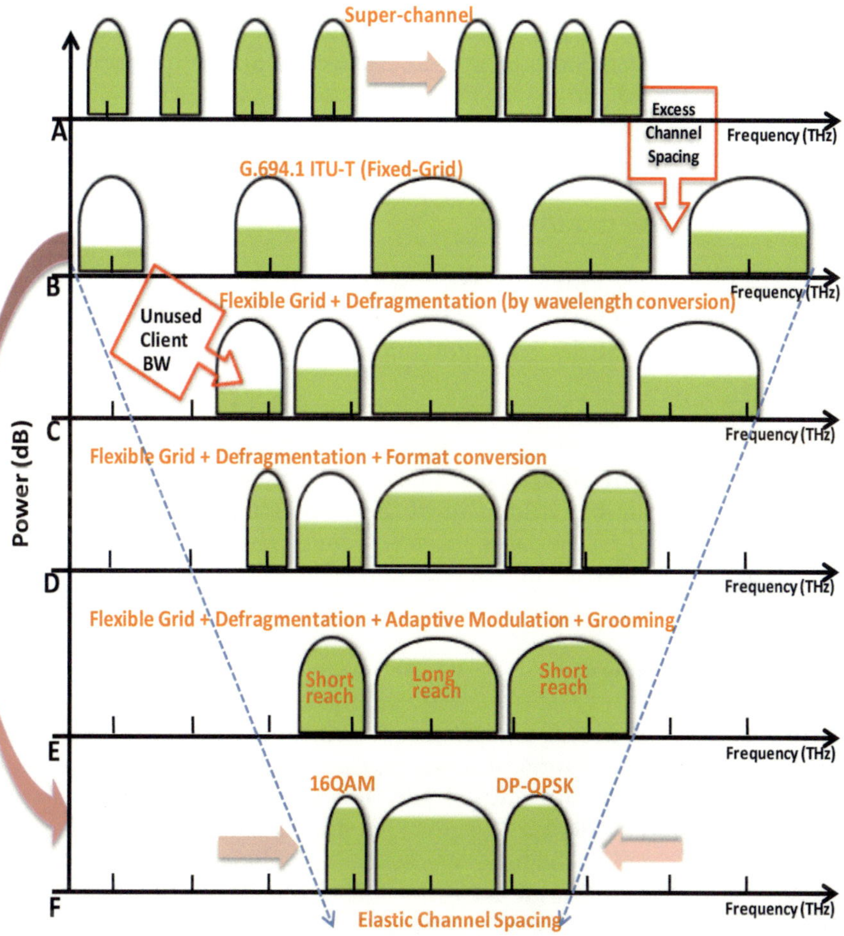

Figure 3.
Roadmap toward elastic spectrum by all-optical signal processing based on transmission distance and user traffic volume [27].

spacing. In order to fully realize the vision of elastic optical networking, network operators have migrated to flexible transport technology solutions capable of supporting grid spacing flexibility. Opportunities for exploiting underutilized spectral resources are shown in **Figure 3**. The essence of elastic optical path network yields highly-efficient optical path accommodation. The resulting spectral savings is achieved by taking advantage of the spectral resources that had not yet been fully utilized, thus results in an increase in network capacity [25]. Let us consider an example in which mixed-rate with different modulation format traffic is transported in a single fiber. For spectrum A, four 100-Gb/s optical channels headed for the same destination. These can be combined them into a tightly spaced 400-Gb/s superchannel and transported as a single entity by eliminating the unnecessary spacing between the channels. Spectrum B shows the ITU-T fixed grid with excess channel spacing. In the next step (spectrum C) implement a flexible grid by using the all optical processing technique i.e. defragmentation. For client traffic that does not fill the entire capacity of a wavelength, the elastic optical path network provides the right sized intermediate bandwidth [26] by format conversion through adaptive modulation represented in spectrum D. This makes the unused client bandwidth available for use. The wavelengths routed in the same direction, grooming is performed as illustrated in spectrum E. Finally in spectrum F, for shorter optical paths, which suffer from less SNR degradation, employment of more spectrally efficient modulation format, such as 16QAM or DP-QPSK, further combined with elastic channel spacing, where the required minimum guard band for wavelength routing is assigned between channels is performed [25]. In this way, elastic optical path networks accommodate a wide range of traffic in a highly spectrally efficient manner [26].

3. Optical switching paradigms

To appreciate the benefits and challenges of all optical networking, it is instructive to review switching in the optical layer. In this section, we first discuss the carrier perspective on the design, functionality and application of switching technology.

Ongoing years have exhibited the constrained versatility of electronic switching changing to acknowledge transport systems. In reaction, all-optical switching/processing has been recognized as a candidate arrangement to empower high-capacity communication within the future. One of the elemental challenges is to proficiently bolster a wide range of traffic configuration, driving the requirement for hardware that is affordable to construct and deploy [28]. The purpose is to identify the key switch technology which can be used in the gridless elastic network scenario.

3.1 Switching technology

With the growth of network traffic due to continue to meet the insatiable consumer demand for transmission bandwidth, the need for more flexibility and better control over network capacity which drives the need for switching in the optical domain is apparent [29]. **Figure 4** shows a potential photonics switch evolution. The switch fabric circuit is the fundamental building block of the next generation optical network, distributing all network traffic from ingress ports to egress ports and also providing the functionality of sharing the spectrum in the time domain with respect to sub wavelength switching. The performance of switch fabrics is very critical in network applications. Efficiencies must be delivered via the use of switch management which is autonomic and guided by intelligent software algorithms.

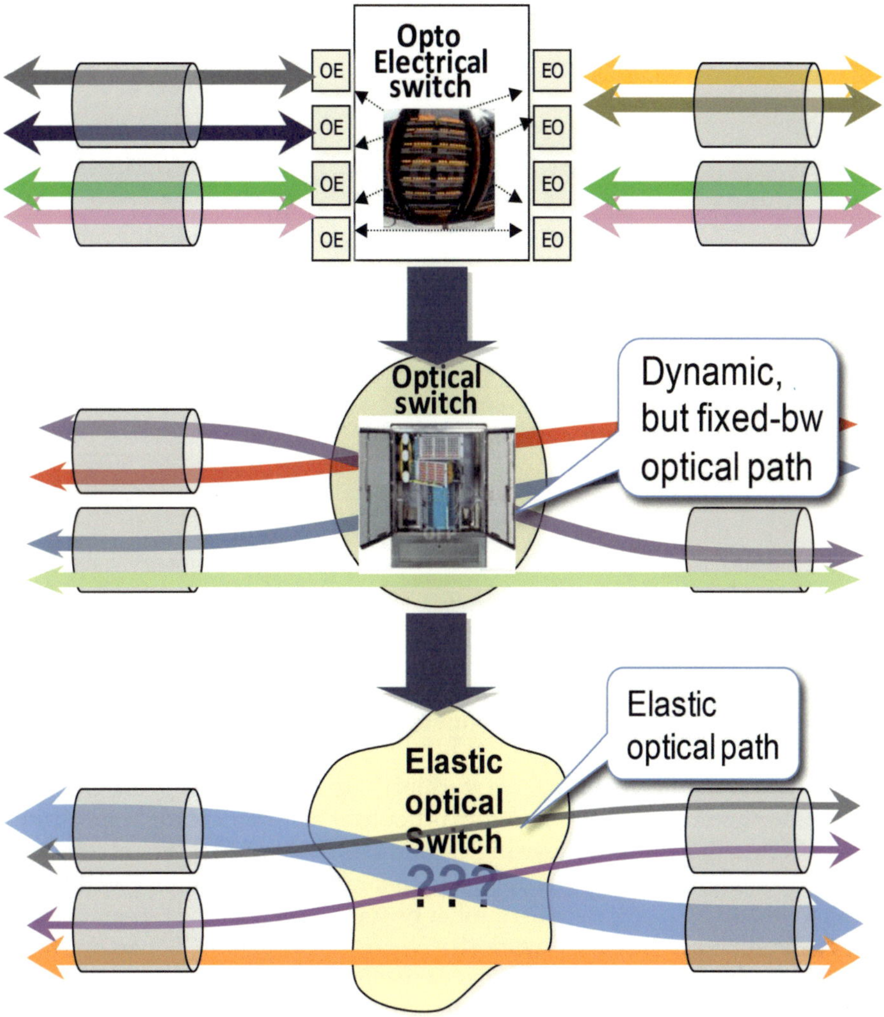

Figure 4.
Photonic switch evolution [27].

As mentioned elastic optical communication are based on the principle that the bandwidth of fiber can be partitioned dynamically into adaptable size spectrum slots. The size and state of each space are generally customized to the prerequisites of a particular (group of) channel(s) in order to achieve effective network-wide transport. This fine slicing and shaping of passbands is achievable with spectrum selective switching (SSS) devices feature a fine spectrum granularity that facilitates the employment of customizable filters with variable bandwidth, e.g. from 10 GHz to 5 THz in 1-GHz increments, and attenuation, e.g. programmable from 4 dB to 30 dB in 1-dB steps [30].

A few optical switch technologies are shown in **Table 1** that can be used in elastic networks. The combination of gridless spectrum switching and rapid time switching devices advances empowers the provisioning of a wide range of optical bandwidth granularities [6]; however, they do not have a large port-count, which makes them incompatible for connecting devices in an elastic optical node. Fast switching are more easily achieved with semiconductor optical amplifiers (SOA) or electro-optic materials such as LiNbO3 or PLZT. Large port-counts are usually implemented with 3D-MEMS [38] or direct beam-steering devices [39] (achieves lower insertion

Material	Cons	Pros
PLZT [31]	Coupling loss	High-speed switching (<5–10 ns), low driving voltage (5–10 V)
LiNbO3 [32]	Polarization dependence	Fast response (<10 ns), low voltage (3 V)
SOA [33]	High PDL, wavelength dependence	Fast response (3 ns), no insertion loss, low power consumption
Directional Coupler [34]	High insertion loss	Fast response (~100 ps), low cross talk
Polymer [35]	Slow response (>1 ms), medium power consumption	Low loss (<0.5 dB/cm)
Silica (PLC) [36]	High power consumption	High extinction ratio
MEMS [37]	Slow response (~10 ms), high voltage (50–200 V)	Scalability, small cross talk

Table 1.
Optical switch technologies.

loss than 3D-MEMS [40]), they have a slower response than fast switching devices, on the order of 10 ms.

4. Need for elastic optical node in optical fiber communication

Recent developments in optical networking have enabled elastic allocation of spectral resources in a gridless manner in order to accommodate high-speed and variable-bitrate signals and achieve high spectral efficiency [41]. For instance, in [42] the spectrum sliced elastic optical path network centered on OFDM, alters signal bandwidths by changing the number of sub-carriers in the transmitted OFDM channel. Other elastic networks based on single carrier (SC) adapt the transmission modulation format in order to mitigate losses or transmit at higher data-rates by exploiting the high OSNR margin [8] with a tradeoff between transmission range and spectral efficiency (SE). Such demonstration requires optical communication infrastructure with the suitable functionality in order to operate [8], frequency/time/space multiplexing, spectrum defragmentation, wavelength conversion, multicasting, format conversion, etc. However, it would be cost-prohibitive to deploy a complete set of technologies and infrastructure to fulfill the requirements for all possible signals and traffic profiles. Instead, emerging technologies need to co-exist with existing ones and provide a smooth migration path where old technology can be gradually replaced. Also, in the context of dynamic optical networks the required services functionalities might change overtime as channels with specific transport and switching requirements are setup and terminated. Providing efficient support for this combination of dynamic requirements with static optical node architectures is a major challenge, which may not be achievable or cost effective [8]. Thus, a new type of flexible and evolvable optical infrastructure needs to be developed in order to enable flexible allocation of resources, and provide any switching and processing capability on demand.

The network scenario considered herewith takes into account the potential applicability of all-optical processing techniques in the network domain. **Figure 5** depicts a case within this framework which consists of four stages which can be placed in diverse geographical sites, i.e. arbitrary input traffic 12.5/42.7/170.8/Gb/s transmitters (WDM domain), an all-optical processing node, field transmission (dark fiber) and receiver. Upon entering the all-optical processing node and to

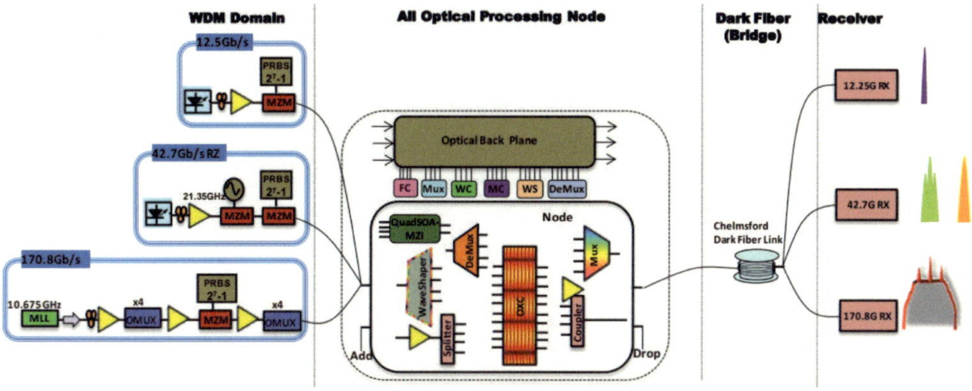

Figure 5.
Network scenario and switching node architecture. Four network sections are highlighted: transmission, all-optical processing in flexible architecture adaptable node (FC, format conversion; WC, wavelength conversion; WS, WaveShaper; MC, multicasting; OXC, optical cross connect), dark fiber, receiver [7].

deliver the optical routing function, all the WDM signals are routed through the 96 × 96 3D-microelectromechanical systems. The optical cross connect architecture consist of subsystems such as Quad Semiconductor Optical Amplifier-Mach-Zehnder interferometer, 200 ms LCoS based SSS [43], wavelength/waveband AWG, (De)-Multiplexer, optical power couplers/splitters and EDFA are interconnected using a 96 × 96 3D-microelectromechanical systems optical switch. The grooming node delivers required real time all-optical processing [44] functionalities, arbitrary spectrum switching and time-domain sub-wavelength switching [45]. It also considerably improves the efficiency and elasticity of the optical node and offers support for current and future data-rates with transparent facilities with low power consumption. All the traffic after node is routed over a dark fiber link, and performances are evaluated by means of BER measurements [7].

5. Key functional building blocks for next generation elastic optical network: wavelength conversion, format conversion and multicasting

The foundation of a photonic network's physical layer is transmission technology in the links and switching technology in the nodes. Different optical multiplexing techniques and modulation formats such as differential quadrature phase shift keying (DQPSK) and quadrature amplitude modulation (QAM) can increase the transmission line rates per fiber to more than 100 Tbit/s [46]. Now research effort are targeted at the Optical Transport Network layer switching where optical processing in the network node allows highly efficient use of capacity, and the abstraction of client data rates from the super-channel data rate. Since some of the networking functions are difficult to carry out electrically, novel processing schemes are required. All-optical processing techniques remove the need for optical-to-electrical conversion, and electronic processing, resulting in optically transparent networks [47]. **Figure 6** illustrates the key issues relating to the component and subsystem requirements for the main parts of the future all optical network [48]. As discussed earlier, flexibility is the important issue that will drive optical subsystem research and development over the next few years.

One of the challenges being faced by the elastic network is bandwidth assignment and channel numbering for different bit rates. When increasing number of channels, the spectrum gets fragmented as channels are added and removed leaving behind noncontiguous empty slots. When high bandwidth requests arrive there may

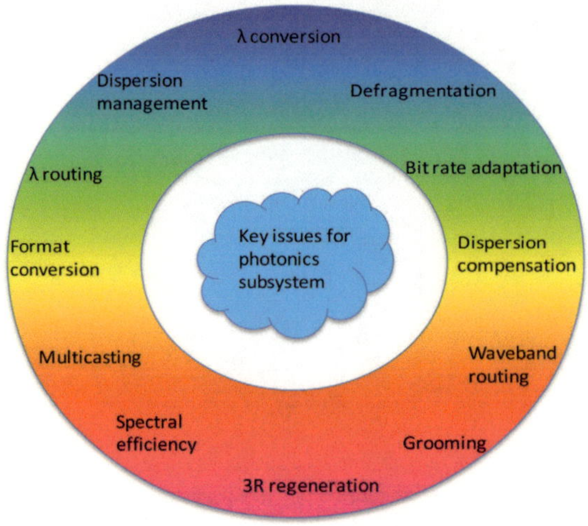

Figure 6.
Key issues of future photonics subsystem [27].

not be sufficient contiguous bandwidth to accommodate them, resulting in blocking [49]. Techniques for spectrum defragmentation involve relocation of existing wavelengths by means of wavelength conversion. Providing such functionality to those channels that require it, and doing it in an efficient manner, is a major challenge. This holds true for other types of functionality such as time multiplexing, format conversion, regeneration, etc. Importantly all-optical modulation format conversion is likely to be used for future all-optical networks in order to add the optical network flexibility [50, 51]. In these networks, systems deployed in different regions could have different bit rates and modulation formats depending on the network size. Therefore, a critical requirement will be the transparent interconnection of these different network islands, which should take place by all-optical means at the network gateways [52].

Transparent optical multicast by multiwavelength conversion has revealed a brand-new way for performing data multicast function directly in the optical domain without passing through any electronics. It provides new visions of optical network designs in terms of optical network switching and forwarding efficiency, transparency, and effectiveness [53]. One-to-many or multichannel wavelength converters are very attractive because they could potentially reduce the number of converters in a routing node without adding more complexity in the switch design. Applications for optical multicast include teleconferencing, video distribution, multiparty gaming and global enterprise virtual private networks (VPNs), etc. [54].

6. All optical converters

The implementation of simple converters can be considered an enabling technology for taking the full advantage of the wavelength dimension in WDM networks. A straightforward implementation of a converter would be a detector followed by an electronic amplifier and a transmitter with the desired new output format/wavelength. However this principle suffers from both a high component count and a high power consumption making the approach impractical for use in large optical switches. Therefore, much attention has been devoted to the realization of all optical converters exploiting the properties of nonlinear devices relying on mechanisms such as cross gain modulation (XGM), cross phase modulation (XPM) and four-wave mixing (FWM) in nonlinear devices [55].

6.1 Cross gain modulation (XGM)

In its general form the principle of operation of the technique is that an optical input signal to be wavelength converted is used to saturate the gain of an active nonlinear element and thereby modulate a continuous wave (CW) signal (pump) at the desired output wavelength [49, 56–58] as shown in **Figure 7**.

Several approaches have emerged to implement all optical wavelength conversion using XGM in SOA, but its conversion speed is determined by the carrier dynamics [59], which are governed by slow interband carrier recombination [60]. Some indicative examples include the work presented in [61] where a 1.2 mm long SOA is used for 40 Gb/s conversion with 1.5 dB power penalty. In [62] a 2 mm long SOA is used for achieving 100 Gb/s wavelength conversion based on XGM. In [63] 80 Gb/s conversion with reportedly low penalty has been achieved.

XGM WCs are attractive because of their simplicity and high conversion efficiency, and the conversion can be made independent of the polarization of the incoming signal. The XGM converter has a number of shortcomings, such as (bit stream) data polarity inversion and the relatively large chirp of the output signal due to the large gain modulation. Finally other than wavelength conversion, XGM has been used for many functions like: format conversion [64], multicasting [65] and header processing in packet switches [66, 67].

6.2 Cross phase modulation (XPM)

In XPM, phase change of the nonlinear element is used rather than the gain change. The optical input signal power controls the phase difference acquired by a pump along the two arms through the refractive index of the nonlinear element [68]. There are number of interferometric configurations that have been used for XPM, some of these are shown in **Figure 8**. A very promising method is XPM in delayed interference signal wavelength convertor configuration (DISC) [62, 69]. 3R regeneration with DISCs has been demonstrated in [70] and a Michelson's interferometer (MI) [71], Mach-Zehnder (MZ) [72]. Some of the first implementations to appear using the interferometric technique is Terahertz Optical Asymmetrical Demultiplexer (TOAD) [73] configuration for the MZI where high-speed operation has been achieved [75].

These converters can perform ultra-high bit rate operations, e.g., demultiplexing from 250 Gb/s [76]. Configurations like the TOAD or the Ultra-fast Nonlinear Interferometer (UNI) [77] have been used mainly for optical processing functions other than wavelength conversion. For instance, UNI configurations have successfully been used in [78] for optical packet switching and in [79] for clock recovery.

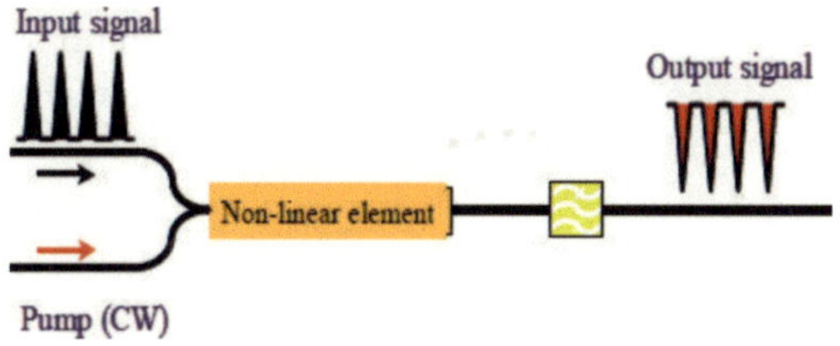

Figure 7.
Schematic of wavelength conversion by XGM in nonlinear medium [27].

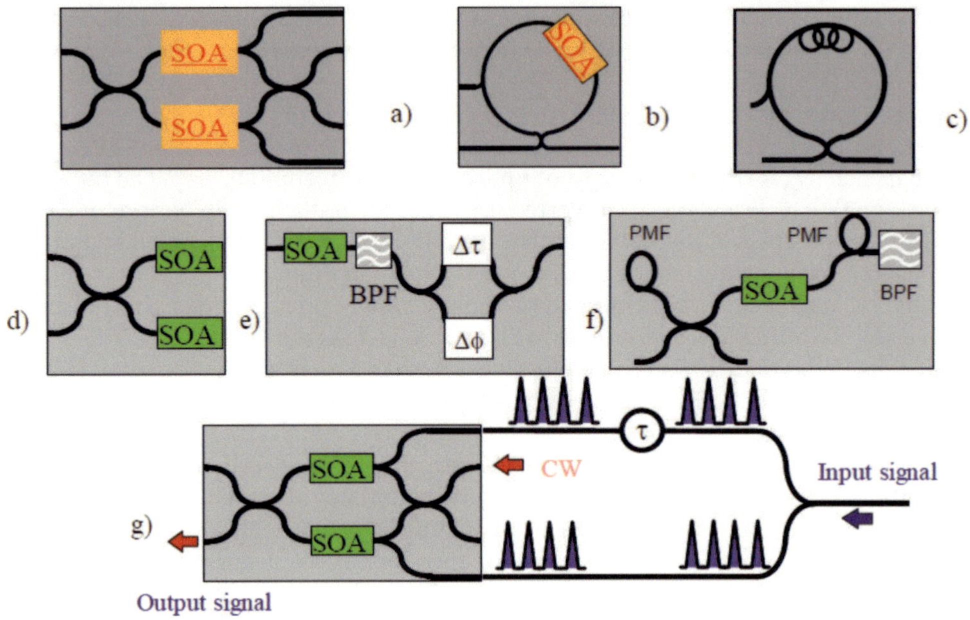

Figure 8.
Different interferometric configurations that have been used as optical gates (a) MZ, (b) TOAD, (c) NOLM, (d) MI, (e) DISC or (f) UNI and (g) push-pull [74].

The use of SOAs is favored over the other elements mentioned above by virtue of compactness and the potential for amplification in suitable all active configurations [56].

The most promising implementation is based on Mach-Zehnder (MZ) structures, their compactness and broad range of functionalities give a very versatile, and thereby a very cost effective device. Nevertheless, owing to the carrier recovery time, the operating bit-rate of the MZ is limited, like the XGM gates. To tackle this problem push-pull configurations have been suggested where, by applying phase changes in both arms [56], the performance can be significantly enhanced [80, 81]. This configuration has been used in a number of different signal processing applications such as demultiplexing [81], regeneration [82], add-and-drop multiplexing [83], regenerative add-and-drop multiplexing [84] and format and wavelength conversion [85]. In fiber-based devices, the nonlinear loop mirror (NOLM) is of particular interest [86, 87]. Due to the inherently ultra-fast response of the Kerr nonlinearity, the NOLM is capable of performing a number of fast bit-level processes e.g., 640 to 10 Gb/s demultiplexing [88], regeneration [89], simultaneous 10 Gb/s wavelength conversion and regeneration [90], and clock extraction [91]. Other than in interferometric configurations, phase modulation can be translated into power modulation by a simple notch filter like that in [92] and XPM can be achieved. XPM in fibers has been used in other configurations to provide 160 Gb/s conversion and 3R regeneration [93] and low penalty wavelength conversion [94]. Recently Raman enhanced self-phase modulation in fibers attracted a lot of attention as an ultra fast technique with noise suppression capabilities [95], however the method is still immature. Generally, nonintegrated devices suffer from stability problems and encounter control issues.

6.3 Wave mixing based: four wave mixing (FWM)

The linearity of the optical reaction is lost when a high-power optical signal is introduced into a fiber. Four wave mixing is a sort of optical Kerr impact, happens when light of two or more unlike wavelengths is launched in a fiber, offering ascend to another wave. When two pump photons are formed, two photons are created: the first one at the signal frequency, the other one at a complementary frequency called idler as shown in **Figure 9** [96, 97]. A schematic of the FWM process, that takes place in nonlinear media, is shown in **Figure 10**. The beating of two distinctive frequency waves modulates the medium's polarization and generates a grating. The input wave interaction with the gratings leads to new components of the frequency. The cause of FWM in SOAs [98–102], is linked to interband and intraband carrier dynamics, while in passive devices and fibers it is because of the induced polarization of the medium under an electric field. The major disadvantage of FWM is its low efficiency, which results in low power FWM products. The main parameter that affects both the efficiency and optical signal-to-noise ratio (OSNR) is the unsaturated gain [68, 103], which can be enhanced by utilizing either longer SOAs, with a smaller active layer [104] or different structures such as multiquantum well devices [103]. The use of an assisted beam has been suggested for the enhancement of the FWM efficiency in [105, 106].

Another disadvantage is that FWM is normally polarization sensitive. The problem has been tackled by using dual-pump configurations [107, 108]. In [109] a 80 nm conversion bandwidth was achieved by dual pump configuration at 2.5 Gb/s. In [110] an almost constant efficiency (<3 dB variation) over a 36 nm wavelength conversion of a 10 Gb/s channel, was achieved.

Due to the nature of the nonlinearities, four wave mixing is very fast. In [111] a multiplexed channel of 100 Gb/s was converted over a range of 3.2 nm by a 2 mm SOA. In addition to converting high bit rate pulses [112–114], the method has been utilized to convert modulation formats [115, 116]. Very fast FWM conversion has been applied as in the demonstration of a 6.3 GHz clock extraction from a 400 Gb/s signal [117], 100 to 10 Gb/s demultipler based on photonic downconversion for a stable add/drop operation [118]. Regenerative properties of wavelength converters based on FWM in a semiconductor optical amplifier have also been demonstrated in [119].

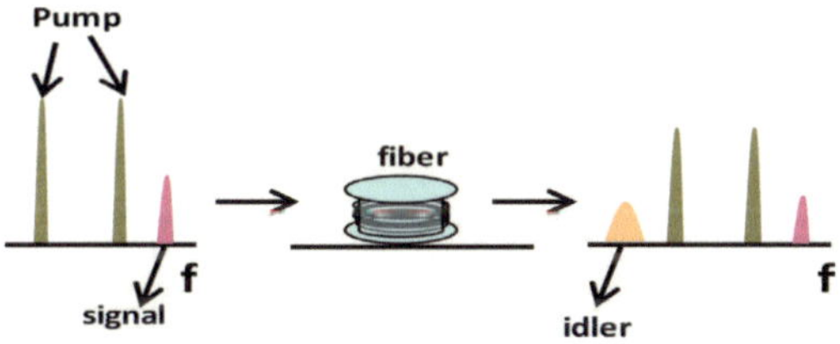

Figure 9.
FWM effect in the spectral domain after the propagation of the signal and the two pumps in DSF fiber [27].

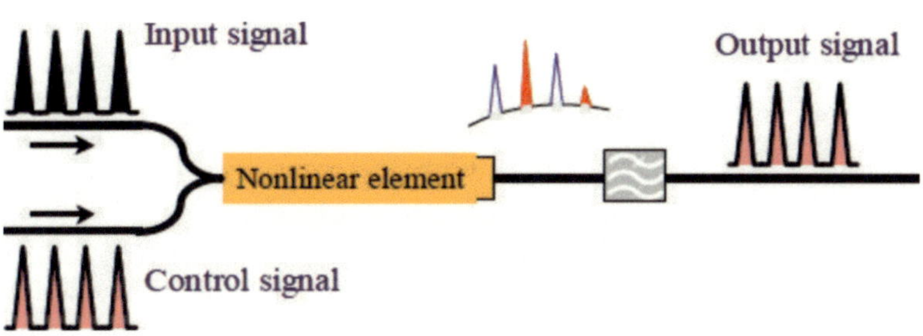

Figure 10.
Mixing of a control and input signal for a four-wave mixing based gate [27].

Another advantage of FWM is that it supports the simultaneous conversion of multiple wavelengths. This has been demonstrated in [48] where 26 WDM channels were simultaneously converted in a highly nonlinear, specially designed fiber. In [101, 120] multichannel SOA based FWM is investigated. In [65, 121] it has been used for multicasting. In a highly nonlinear fiber for FWM was used to demonstrate simultaneous conversion of 200 Gb/s [122] and 32 × 10 Gb/s channels [123].

FWM is a very promising technique, but due to the complexity of the configuration for polarization insensitive operation or wide tunability it will probably be used only for converters operating at 100 Gb/s and beyond.

7. All optical signal processing functionalities survey

Many efforts have been put in the past years to design and implement the entire collection of subsystems required for enabling high speed all-optical routing and switching by all-optical signal processing exploiting a variety of nonlinear media [124]. Significant examples are, format conversion by cross phase modulation at 40 Gb/s based on single SOAs [125] and 160 Gb/s all-optical OOK to DPSK format converter based on HNLF [126]. Both the schemes require high input power, i.e. high clock power in the first case and high input power, 20 dBm for the HNLF, making their operation expensive in terms of energy consumption. Coupling loss of 8 dB between the lensed fiber and the waveguide is reported in [127], as reverse biased silicon-on-isolator p-i-n rib waveguides is wavelength converted via FWM. In [128], demonstration of FC NRZ-DPSK-to-RZ-DPSK, NRZ-DPSK-to-RZ-DPSK and NRZ-to-RZ in a PPLN waveguide by using SHG/DFG, while all-optical wavelength MC based on combination of XGM and XPM in a QD-SOA is presented in [129]. Here the detrimental effect of FWM coherent cross talk on the conversion is identified as the major drawback, responsible for degradation when using four equally spaced channels. Recently, T-H Cheng et al. demonstrated the multicasting in a HNLF loop mirror using FWM at 10 Gbit/s, but using three-pump lasers. In [130], demonstration of Grooming Switch for an OTDM Meshed Networking was reported, however OTDM-to-WDM would need thermally stabilized packaging and the light-paths in the WDM-to-OTDM must preferably be made of waveguides rather than fiber. Each of the demonstration could not support simultaneous MC, FC and WC of multiple modulation format signals to groom OTDM signals in a gridless elastic communication [7]. **Table 2** summarizes the devices used in all optical processing functionalities to date.

The suitability of optical wavelength converters for future networks will be judged on specific criteria that these must fulfill [56, 60]. In particular, modules will ideally have to simultaneously present the following characteristics: compactness (can be integrated in a single substrate with the other switch modules), operation at low optical/electrical powers with high dynamic range, polarization insensitivity, complete transparency to bit-rate (>100 Gb/s) and format or easily adjustable, induce minimal transmission power penalty (small chirp, amplitude distortion and extinction ratio reduction, and large OSNR) to a signal and thus can be cascaded, provision of amplification and (ideally) regeneration and wide conversion bandwidth (tunability) without the need of filtering.

The fundamental idea, common for all technologies discussed here, is the exploitation of the physical properties of a nonlinear element to perform optical processing. The main nonlinear elements are: SOA-MZI, PPLN and HNLF. SOA based devices have the added advantages of compactness and low energy requirements to trigger nonlinearities. Fibers have an instantaneous response to pulses but on the other hand have limited nonlinearity, even in specially designed photonic crystal fibers, hence long lengths and high injected powers are required for efficient operation [74].

Reference	Device	Reference	Device
[131]	SOA 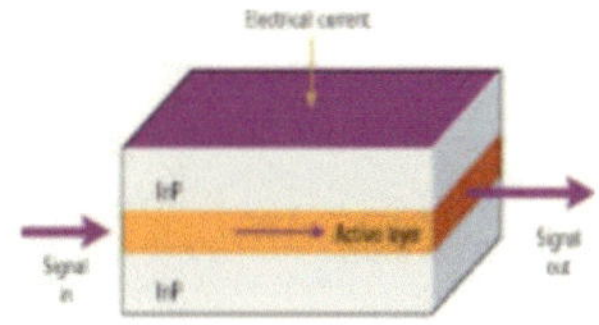	[132]	Adhered ridge waveguide 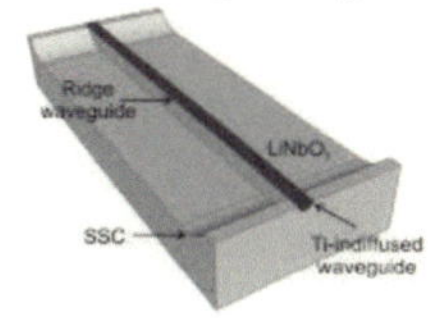
[133]	PPLN 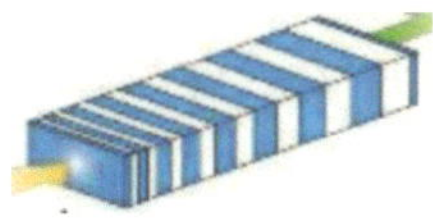	[134]	SOA-MZI 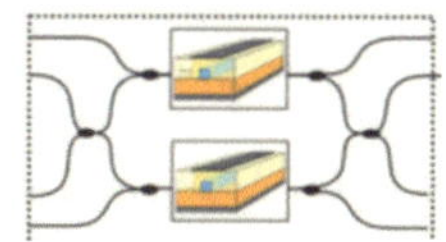
[135]	Chalcogenide microstructured fiber 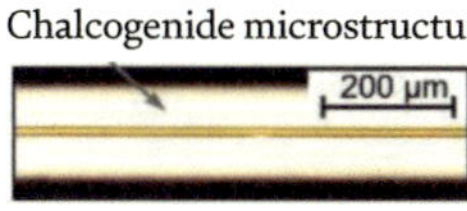	[136]	HNLF 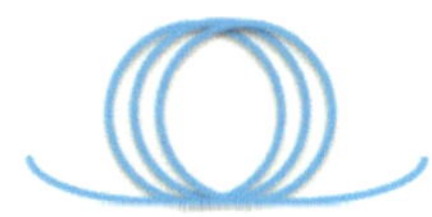
[31]	Silicon nanowaveguides 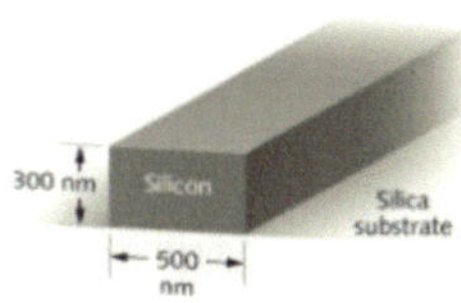	[137]	ENZ 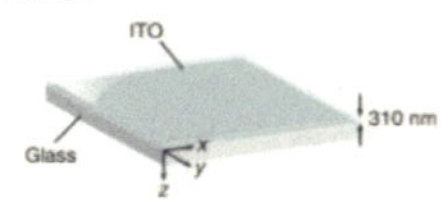

Table 2.
Major technologies devices for all-optical processing reported to date.

8. Conclusion

Exploring methods for the processing of signals in the optical domain, the chapter includes the solutions for the efficient transport of heterogeneous traffic by enhancing the flexibility of the optical layer in allocating network resources as well as for the implementation of an adaptable infrastructure that provides on-demand functionality according to traffic requirements. Further provides a comprehensive review of the state-of-the art of optical signal processing technologies and devices. It presents breakthrough solutions for enabling a pervasive use of optical signal processing to overcome the capacity crunch in optical fiber applications. The chapter content ranges from the road map toward elastic optical networks, optical switching paradigms, need for elastic optical node and application that support gridless node in optical communication having the ability of repositioning signals in a fragmented spectrum by all-optical signal processing functionalities such as MC, spectrum defragmentation, FC, WC and grooming of high speed signals in order to maintain a proficient resource utilization.

Author details

Muhammad Irfan Anis
Iqra University, Karachi, Pakistan

*Address all correspondence to: mirfananis@iqra.edu.pk

References

[1] Yan L, Willner AE, Wu X, Yi A, Bogoni A, Chen Z-Y, et al. All-optical signal processing for ultrahigh speed optical systems and networks. Journal of Lightwave Technology. 2012;**30**(24):3760-3770

[2] Willner AE, Fallahpour A, Alishahi F, Cao Y, Mohajerin-Ariaei A, Almaiman A, et al. All-optical signal processing techniques for flexible networks. Journal of Lightwave Technology. 2018;**37**(1):21-35

[3] Ji Y, Zhang J, Zhao Y, Yu X, Zhang J, Chen X. Prospects and research issues in multi-dimensional all optical networks. Science China Information Sciences. 2016;**59**(10):101301

[4] Willner AE, Khaleghi S, Chitgarha MR, Yilmaz OF. All-optical signal processing. Journal of Lightwave Technology. 2013;**32**(4):660-680

[5] Zhong K, Zhou X, Huo J, Yu C, Lu C, Lau APT. Digital signal processing for short-reach optical communications: A review of current technologies and future trends. Journal of Lightwave Technology. 2018;**36**(2):377-400

[6] Amaya N, Zervas GS, Rofoee BR, Irfan M, Qin Y, Simeonidou D. Field trial of a 1.5 Tb/s adaptive and gridless OXC supporting elastic 1000-fold all-optical bandwidth granularity. Optics Express. 2011;**19**(26):B235-B241

[7] Anis MI, Amaya N, Zervas G, Pinna S, Scaffardi M, Fresi F, et al. Field trial demonstration of spectrum defragmentation and grooming in elastic optical node. Journal of Lightwave Technology. 2013;**31**(12):1845-1855

[8] Jinno M, Takara H, Kozicki B, Tsukishima Y, Sone Y, Matsuoka S. Spectrum-efficient and scalable elastic optical path network: architecture, benefits, and enabling technologies. IEEE communications magazine. 2009;**47**(11):66-73

[9] Jensen R, Lord A, Parsons N. Colourless, directionless,contentionless ROADM architecture using low-loss opticalmatrix switches. In: Proc. European Conf. Optical Communication (ECOC), Torino, Italy, Paper Mo.2.D.2. 2010. pp. 1-3

[10] Wen K, Yin Y, Geisler DJ, Chang S, Yoo SJB. Dynamic on-demand lightpath provisioning using spectral defragmentation in flexible bandwidth networks. In: European Conference and Exposition on Optical Communications. Geneva, Switzerland: Optical Society of America; 2011. p. Mo-2

[11] Alabarce MG. Politecnico di Torino. 2013. Available: https://iris.polito.it/retrieve/handle/11583/2506247/59343/phd_thesis_Garrich.pdf

[12] Winzer PJ et al. Generation and 1,200-km transmission of 448-Gb/s ETDM 56-Gbaud PDM 16-QAM using a single I/Q modulator. In: 2010 36th European Conference and Exhibition on Optical Communication (ECOC), Torino, Italy, Paper PDP2. 2. 2010. pp. 1-3

[13] Chandrasekhar S, Liu X, Zhu B, Peckham DW. Transmission of a 1.2-Tb/s 24-carrier no-guard-interval coherent OFDM superchannel over 7200-km of ultra-large-area fiber. In: Proceedings of European Conference on Optical Communication (ECOC), Vienna, Austria, Paper PD2.6. 2009. pp. 1-3

[14] Hillerkuss D, Schellinger T, Schmogrow R, Winter M, Vallaitis T, Bonk R, et al. Single source optical OFDM transmitter and optical FFT receiver demonstrated at line rates of

5.4 and 10.8 Tbit/s. In: Optical Fiber Communication Conference; San Diego, CA. Optical Society of America; 2010. p. PDPC1

[15] Yu J, Dong Z, Chi N. Generation, transmission and coherent detection of 11.2 Tb/s (112 × 100 Gb/s) single source optical OFDM superchannel. In: Optical Fiber Communication Conference; Los Angeles, CA. Optical Society of America; 2011. p. PDPA6

[16] Huang Y-K, Ip E, Xia TJ, Wellbrock GA, Huang M-F, Aono Y, et al. Mixed line-rate transmission (112-Gb/s, 450-Gb/s, and 1.15-Tb/s) over 3560 km of field-installed fiber with filterless coherent receiver. Journal of Lightwave Technology. 2011;**30**(4):609-617

[17] Eiselt M, Teipen B, Grobe K, Elbers J-P. WDM transport towards Terabits/s line rates—What will be gained? In: Photonic Networks, 12. ITG Symposium. Leipzig, Germany: VDE; 2011. pp. 1-6

[18] Kakande J, Slavík R, Parmigiani F, Petropoulos P, Richardson D. All-optical processing of multi-level phase shift keyed signals. In: Optical Fiber Communication Conference; Los Angeles, CA. Optical Society of America; 2012. p. OW1I-3

[19] Payne DN. The optical fibre Internet: Where next? In: ICTON'12. Coventry, UK: University of Warwick; 2012

[20] Gringeri S, Basch B, Shukla V, Egorov R, Xia TJ. Flexible architectures for optical transport nodes and networks. IEEE Communications Magazine. 2010;**48**(7):40-50

[21] Angelou ITEPM. A survey of recent developments on flexible/elastic optical networking. In: ICTON'12. Coventry, United Kingdom: University of Warwick; 2012

[22] Miyamoto Y, Kawamura R. Space division multiplexing optical transmission technology to support the evolution of high-capacity optical transport networks. NTT Technical Review. 2017;**15**(6):1-7

[23] Essiambre RJ, Kramer G, Winzer PJ, Foschini GJ, Goebel B. Capacity limits of optical fiber networks. Journal of Lightwave Technology. 2010;**28**(4):662-701

[24] Ip EJ, Mateo P, Huang E, Xu YK, Qian L, Bai D, et al. 100G and beyond transmission technologies for evolving optical networks and relevant physical-layer issues. Proceedings of the IEEE. 2012;**100**(5):1065-1078

[25] Jinno M, Takara H, Sone Y, Yonenaga K, Hirano A. Elastic optical path network architecture: Framework for spectrally-efficient and scalable future optical networks. IEICE Transactions on Communications. 2012;**95**(3):706-713

[26] Jinno M, Takara H, Yonenaga K. Why do we need elastic optical path networking in the 1 Tb/s era? In: Conference on Lasers and Electro-Optics/Pacific Rim; Sydney, Australia. Optical Society of America; 2011. p. C74

[27] Anis MI. An experimental investigation of all-optical signal processing techniques for application in elastic optical networking [PhD]. UK: School of Computer Science and Electronic Engineering, University of Essex; 2014

[28] De Leenheer M, Develder C, Buysse J, Dhoedt B, Demeester P. Performance analysis and dimensioning of multi-granular optical networks. Optical Switching and Networking. 2009;**6**(2):88-98

[29] Mahony MO, Politi C, Hill G. Roadmap on optical transport network technologies. In: 2008 10th

Anniversary International Conference on Transparent Optical Networks, Athens, Greece. Vol. 1. IEEE; 2008. pp. 165-168

[30] Guide P. WaveShaper—Family of Programmable Optical Processors. Available from: https://webcache.googleusercontent.com/search?q=cache:tFoPyk7UtlYJ:https://www.digikey.com/Site/Global/Layouts/DownloadPdf.ashx%3FpdfUrl%3DE470BF5108FB43B9A011925D65AB41DD+&cd=1&hl=en&ct=clnk&gl=pk&client=firefox-b-d

[31] Wang Z, Liu H, Sun Q, Huang N, Han J. All-optical wavelength conversion based on four-wave mixing in dispersion-engineered silicon nanowaveguides. Journal of Russian Laser Research. 2017;**38**(2):204-210

[32] Moralis-Pegios M, Terzenidis N, Mourgias-Alexandris G, Vyrsokinos K, Pleros N. A 1024-port optical uni- and multicast packet switch fabric. Journal of Lightwave Technology. 2019;**37**(4):1415-1423

[33] Raja A, Mukherjee K, Roy J, Maji K. Analysis of polarization encoded optical switch implementing cross polarization modulation effect in semiconductor optical amplifier. Probe. 2019;**5**(1):1-5

[34] Barral D, Bencheikh K, Olver PJ, Belabas N, Levenson JA. Symmetry-based analytical solutions to the $\chi(2)$ nonlinear directional coupler. arXiv preprint arXiv:1901.04897. 2019

[35] Ma Y, Sikdar D, Fedosyuk A, Velleman L, Zhao M, Tang L, et al. An auxetic thermo-responsive nanoplasmonic optical switch. ACS Applied Materials & Interfaces. 2019;**11**:22754-22760

[36] Sun Q, Sun DG. Investigation for the thermal conduction of silicon oxide waveguide optical switch. In: 2018 IEEE International Conference on Manipulation, Manufacturing and Measurement on the Nanoscale (3M-NANO), Hangzhou, China. IEEE; 2018. pp. 159-162

[37] Kwon K, Seok TJ, Henriksson J, Luo J, Ochikubo L, Jacobs J, et al. 128× 128 silicon photonic MEMS switch with scalable row/column addressing. In: CLEO: Science and Innovations; San Jose, CA, USA. Optical Society of America; 2018. p. SF1A-4

[38] Nashimoto K, Kudzuma D, Han H. High-speed switching and filtering using PLZT waveguide devices. In: OECC 2010 Technical Digest, Sapporo, Japan. IEEE; 2010. pp. 540-542

[39] Wu MC, Solgaard O, Ford JE. Optical MEMS for lightwave communication. Journal of Lightwave Technology. 2007;**24**(12):4433-4454

[40] Bitting M. New optical switches enable automated testing with true flexibility. In: Proceedings AUTOTESTCON 2004, San Antonio, TX, USA. IEEE; 2004. pp. 361-366

[41] Amaya N, Zervas GS, Simeonidou D. Optical node architectures for elastic networks: From static to architecture on demand. In: 2012 14th International Conference on Transparent Optical Networks (ICTON), Coventry, United Kingdom. IEEE; 2012. pp. 1-4

[42] Li Z, Yu X, Xie S, Wang Y, Wang Y, Zhao Y, et al. Module selection algorithm based on WSS/SSS-hybrid AoD node in dynamic elastic optical networks. In: International Conference on Communications and Networking in China. Cham: Springer; 2018. pp. 194-203

[43] Amaya N, Zervas GS, Simeonidou D. Architecture on demand for transparent optical networks. In: 2011 13th International Conference on Transparent Optical Networks, Stockholm, Sweden. IEEE; 2011. pp. 1-4

[44] Simeonidou D, Amaya N, Zervas G. Infrastructure and architectures on demand for flexible and elastic optical networks. In: European Conference and Exhibition on Optical Communication; Amsterdam, Netherland. Optical Society of America; 2012. p. Tu-3

[45] Anis MI, Amaya N, Zervas GS, Nejabati R, Simeonidou D, Scaffardi M, et al. Defragmentation and grooming on 85.4 Gb/s by simultaneous format and wavelength conversion in an integrated quad SOA-MZI. In: 2012 16th International Conference on Optical Network Design and Modelling (ONDM). Colchester, UK: IEEE; 2012. pp. 1-6

[46] Amaya N, Irfan M, Zervas G, Banias K, Garrich M, Henning I, et al. Gridless optical networking field trial: Flexible spectrum switching, defragmentation and transport of 10G/40G/100G/555G over 620-km field fiber. Optics express. 2011;**19**(26):B277-B282

[47] Amaya N, Zervas G, Irfan M, Zhou Y, Lord A, Simeonidou D. Experimental demonstration of gridless spectrum and time optical switching. Optics Express. 2011;**19**(12):11182-11188

[48] Wada N, Furukawa H. Evolution of dynamic optical networks. In: Presented at the 16th International Conference on Optical Network Design and Modelling (ONDM); Colchester, UK. 2012

[49] Bondarczuk K. All-optical processing for terabit/s wavelength division multiplexed systems using two-photon absorption in a semiconductor micro-cavity [doctor of philosophy]. School of Electronic Engineering Dublin City University; 2009. Available from: http://doras.dcu.ie/3621/2/Thesis_KBondarczuk_3_copyright_clear.pdf

[50] Aso O, Arai S, Yagi T, Tadakuma M, Suzuki Y, Namiki S. Broadband four-wave mixing generation in short optical fibres. Electronics Letters. 2002;**36**(8):709-711

[51] Tzanakaki A. Tunable wavelength conversion in optical networks [PhD]. United Kingdom: University of Essex; 2000

[52] Jun S, Park K, Kim H, Chung H, Lee J, Chung Y. Passive Optical NRZ to-RZ Converter. Vol. 2. Los Angeles, CA: IEEE; 2004. p. 3

[53] Wei X, Leuthold J, Zhang L. Delay Interferometer-based Optical Pulse Generator. Vol. 1. Los Angeles, CA: IEEE; 2004. p. 773

[54] Mishina K, Maruta A, Mitani S, Miyahara T, Ishida K, Shimizu K, et al. NRZ-OOK-to-RZ-BPSK modulation-format conversion using SOA-MZI wavelength converter. Journal of Lightwave Technology. 2006;**24**(10):3751-3758

[55] Rouskas GN. Optical layer multicast: rationale, building blocks, and challenges. Network, IEEE. 2003;**17**(1):60-65

[56] Yan N, Teixeira A, Silveira T, Koonen T, Tafur Monroy I, Monteiro P, et al. Simultaneous multi-wavelength signal conversion for transparent optical multicast. In: European Conference on Networks Optical communications (NOC). 2005

[57] Mikkelsen B. Optical amplifiers and their system applications [PhD]. Electromagnetic Systems, Technical University of Denmark; 1994

[58] Stubkjaer KE. Semiconductor optical amplifier-based all-optical gates for high-speed optical processing. IEEE Journal of Selected Topics in Quantum Electronics. 2002;**6**(6):1428-1435

[59] White I, Penty R, Webster M, Chai YJ, Wonfor A, Shahkooh S. Wavelength switching components for future

photonic networks. Communications Magazine, IEEE. 2002;**40**(9):74-81

[60] Yoo SJB. Wavelength conversion technologies for WDM network applications. Journal of Lightwave Technology. 1996;**14**(6):955-966

[61] Durhuus T, Mikkelsen B, Joergensen C, Danielsen L, Stubkjaer KE. All-optical wavelength conversion by semiconductor optical amplifiers. Journal of Lightwave Technology. 1996;**14**(6):942-954

[62] Danielsen S, Joergensen C, Vaa M, Mikkelsen B, Stubkjaer K, Doussiere P, et al. Bit error rate assessment of 40 Gbit/s all-optical polarisation independent wavelength converter. Electronics Letters. 2002;**32**(18):1688

[63] Leuthold J, Joyner C, Mikkelsen B, Raybon G, Pleumeekers J, Miller B, et al. 100 Gbit/s all-optical wavelength conversion with integrated SOA delayed-interference configuration. Electronics Letters. 2002;**36**(13):1129-1130

[64] Nesset D, Kelly T, Marcenac D. All-optical wavelength conversion using SOA nonlinearities. Communications Magazine, IEEE. 2002;**36**(12):56-61

[65] Norte D, Willner AE. All-optical data format conversions and reconversions between the wavelength and time domains for dynamically reconfigurable WDM networks. Journal of Lightwave Technology. 2002;**14**(6):1170-1182

[66] Xu L, Perros HG, Rouskas G. Techniques for optical packet switching and optical burst switching. Communications Magazine, IEEE. 2002;**39**(1):136-142

[67] Park E, Norte D, Willner A. Simultaneous all-optical packet-header replacement and wavelength shifting for a dynamically-reconfigurable WDM network. Photonics Technology Letters, IEEE. 2002;**7**(7):810-812

[68] Pinto A. Optical networks: A practical perspective. Journal of Optical Networking. 2002;**1**(6):219-220

[69] Vlachos K, Raffaelli C, Aleksic S, Andriolli N, Apostolopoulos D, Avramopoulos H, et al. Photonics in switching: enabling technologies and subsystem design. Journal of Optical Networking. 2009;**8**(5):404-428

[70] Nakamura S, Ueno Y, Tajima K. 168-Gb/s all-optical wavelength conversion with a symmetric-Mach-Zehnder-type switch. Photonics Technology Letters, IEEE. 2002;**13**(10):1091-1093

[71] Leuthold J, Raybon G, Su Y, Essiambre R, Cabot S, Jaques J, et al. 40 Gbit/s transmission and cascaded all-optical wavelength conversion over 1000000 km. Electronics Letters. 2002;**38**(16):890-892

[72] Pedersen R, Nissov N, Mikkelsen B, Poulsen H, Stubkjaer K, Gustavsson M, et al. Transmission through a cascade of 10 all-optical interferometric wavelength converter spans at 10 Gbit/s. Electronics Letters. 2002;**32**(11):1034-1035

[73] Durhuus T, Joergensen C, Mikkelsen B, Pedersen R, Stubkjaer K. All optical wavelength conversion by SOA's in a Mach-Zehnder configuration. Photonics Technology Letters, IEEE. 2002;**6**(1):53-55

[74] Sokoloff J, Prucnal P, Glesk I, Kane M. A terahertz optical asymmetric demultiplexer (TOAD). Photonics Technology Letters, IEEE. 2002;**5**(7):787-790

[75] Politi C, Matrakidis C, Stavdas A. Optical wavelength and waveband converters. In: 2006 International Conference on Transparent Optical Networks. Vol. 1. IEEE; 2006. pp. 179-182

[76] Politi C, Matrakidis C, Stavdas A. Optical wavelength and waveband converters. In: 2006 International Conference on Transparent Optical Networks. Vol. 1. Nottingham, UK: IEEE; 2006

[77] Glesk I, Sokoloff J, Prucnal P. Demonstration of all-optical demultiplexing of TDM data at 250 Gbit/s. Electronics Letters. 2002;**30**(4):339-341

[78] Patel N, Rauschenbach K, Hall K. 40-Gb/s demultiplexing using an ultrafast nonlinear interferometer (UNI). Photonics Technology Letters, IEEE. 2002;**8**(12):1695-1697

[79] Theophilopoulos G, Kalyvas M, Bintjas C, Pleros N, Yiannopoulos K, Stavdas A, et al. Optically addressable 2 x 2 exchange/bypass packet switch. IEEE Photonics Technology Letters. 2002;**14**(7):998-1000

[80] Yiannopoulos K, Pleros N, Bintjas C, Kalyvas M, Theophilopoulos G, Avramopoulos H, et al. All-Optical Packet Clock Recovery Circuit. Vol. 2. Copenhagen, Denmark: IEEE; 2006. pp. 1-2

[81] Wang Q, Zhu G, Chen H, Jaques J, Leuthold J, Piccirilli AB, et al. Study of all-optical XOR using Mach-Zehnder interferometer and differential scheme. IEEE Journal of Quantum Electronics. 2004;**40**(6):703-710

[82] Tajima K. All-optical switch with switch-off time unrestricted by carrier lifetime. Japanese Journal of Applied Physics. 1993;**32**:L1746-L1749

[83] Dutta AK, Dutta NK, Fujiwara M. WDM Technologies: Passive Optical Components. Vol. 1. San Diago, US: Academic Press; 2003

[84] Hess R, Duelk M, Vogt W, Gamper E, Gini E, Besse P, et al. Simultaneous all-optical add and drop multiplexing of 40 Gbit/s OTDM signals using monolithically integrated Mach-Zehnder interferometer. Electronics Letters. 2002;**34**(6):579-580

[85] Fischer S, Dulk M, Gamper E, Vogt W, Hunziker W, Gini E, et al. All-optical regenerative OTDM add-drop multiplexing at 40 Gb/s using monolithic InP Mach-Zehnder interferometer. Photonics Technology Letters, IEEE. 2000;**12**(3):335-337

[86] Jepsen K, Clausen A, Mikkelsen B, Poulsen H, Stubkjaer K. Alloptical Network Interface for Bit Synchronisation and Regeneration. Vol. 5. Edinburgh, UK: IET; 2002. pp. 89-92

[87] Yu J, Zheng X, Peucheret C, Clausen AT, Poulsen HN, Jeppesen P. 40-Gb/s all-optical wavelength conversion based on a nonlinear optical loop mirror. Journal of Lightwave Technology. 2002;**18**(7):1001-1006

[88] Blow K, Doran NJ, Nelson B. Demonstration of the nonlinear fibre loop mirror as an ultrafast all-optical demultiplexer. Electronics Letters. 2008;**26**(14):962-964

[89] Yamamoto T, Yoshida E, Nakazawa M. Ultrafast nonlinear optical loop mirror for demultiplexing 640 Gbit/s TDM signals. Electronics Letters. 2002;**34**(10):1013-1014

[90] Lucek J, Smith K. All-optical signal regenerator. Optics letters. 1993;**18**(15):1226-1228

[91] Yu J, Zheng X, Liu F, Buxens A, Jeppesen P. Simultaneous realization wavelength conversion and signal regeneration using a nonlinear optical loop mirror. Optics Communications. 2000;**175**(1-3):173-177

[92] Nakazawa M, Suzuki K, Yamada E. NOLM oscillator and its injection locking technique for timing clock

extraction and demultiplexing. Electronics Letters. 1996;**32**:1122

[93] Olsson BE, Ohlen P, Rau L, Blumenthal DJ. A simple and robust 40-Gb/s wavelength converter using fiber cross-phase modulation and optical filtering. Photonics Technology Letters, IEEE. 2002;**12**(7):846-848

[94] Schubert C, Ludwig R, Watanabe S, Futami E, Schmidt C, Berger J, et al. 160 Gbit/s wavelength converter with 3R-regenerating capability. Electronics Letters. 2002;**38**(16):903-904

[95] Rau L, Wang W, Camatel S, Poulsen H, Blumenthal DJ. All-optical 160-Gb/s phase reconstructing wavelength conversion using cross-phase modulation (XPM) in dispersion-shifted fiber. Photonics Technology Letters, IEEE. 2004;**16**(11):2520-2522

[96] Galili M, Oxenløwe LK, Zibar D, Clausen A, Jeppesen P. 160 Gb/s Raman assisted SPM Wavelength converter. In: 30th European Conference on Optical Communication. 2004

[97] Osamu ASO, Tadakuma M, Namiki S. Four-wave mixing in optical fibers and its applications. Furukawa Electric Review, Japan. 2000;**105**:46-51

[98] Haris M. Advanced modulation formats for high-bit-rate optical networks [doctor of philosophy]. Electrical and Computer Engineering, Georgia Institute of Technology; 2008

[99] Yamawaku J, Yamazaki E, Takada A, Morioka T. Field trial of virtual-grouped-wavelength-path switching with QPM-LN waveband converter and PLC matrix switch in JGN II test bed. Electronics Letters. 2005;**41**(2):88-89

[100] Guekos G. Photonic Devices for Telecommunications. Germany, Berlin: Springer-Verlag; 1999. pp. 269-368

[101] Diez S, Schmidt C, Ludwig R, Weber HG, Obermann K, Kindt S, et al. Four-wave mixing in semiconductor optical amplifiers for frequency conversion and fast optical switching. IEEE Journal of Selected Topics in Quantum Electronics. 2002;**3**(5):1131-1145

[102] Geraghty DF, Lee RB, Verdiell M, Ziari M, Mathur A, Vahala KJ. Wavelength conversion for WDM communication systems using four-wave mixing in semiconductor optical amplifiers. IEEE Journal of Selected Topics in Quantum Electronics. 2002;**3**(5):1146-1155

[103] Hsu A, Chuang S. Wavelength conversion by dual-pump four-wave mixing in an integrated laser modulator. Photonics Technology Letters, IEEE. 2003;**15**(8):1120-1122

[104] D'ottavi A, Girardin F, Graziani L, Martelli F, Spano P, Mecozzi A, et al. Four-wave mixing in semiconductor optical amplifiers: A practical tool for wavelength conversion. IEEE Journal of Selected Topics in Quantum Electronics. 2002;**3**(2):522-528

[105] Girardin F, Eckner J, Guekos G, Dall'Ara R, Mecozzi A, D'Ottavi A, et al. Low-noise and very high-efficiency four-wave mixing in 1.5-mm-long semiconductor optical amplifiers. Photonics Technology Letters, IEEE. 1997;**9**(6):746-748

[106] Lee SL, Gong PM, Yang CT. Performance enhancement on SOA-based four-wave-mixing wavelength conversion using an assisted beam. Photonics Technology Letters, IEEE. 2002;**14**(12):1713-1715

[107] Lee SL, Gong PM, Lin YM, Lee SSW, Yuang MC. High-efficiency wide-band SOA-based wavelength converters by using dual-pumped four-wave mixing and an assist beam.

Photonics Technology Letters, IEEE. 2004;**16**(8):1903-1905

[108] Hasegawa T, Inoue K, Oda K. Polarization independent frequency conversion by fiber four-wave mixing with a polarization diversity technique. Photonics Technology Letters, IEEE. 2002;**5**(8):947-949

[109] Morgan TJ, Tucker RS, Lacey JPR. All-optical wavelength translation over 80 nm at 2.5 Gb/s using four-wave mixing in a semiconductor optical amplifier. Photonics Technology Letters, IEEE. 2002;**11**(8):982-984

[110] Mak MWK, Tsang H, Chan K. Widely tunable polarization-independent all-optical wavelength converter using a semiconductor optical amplifier. Photonics Technology Letters, IEEE. 2000;**12**(5):525-527

[111] Kelly A, Ellis A, Nesset D, Kashyap R, Moodie D. 100 Gbit/s wavelength conversion using FWM in an MQW semiconductor optical amplifier. Electronics Letters. 2002;**34**(20):1955-1956

[112] Kelly A, Marcenac D, Nesset D. 40 Gbit/s wavelength conversion over 24.6 nm using FWM in a semiconductor optical amplifier with an optimised MQW active region. Electronics Letters. 2002;**33**(25):2123-2124

[113] Ludwig R, Raybon G. BER measurements of frequency converted signals using four-wave mixing in a semiconductor laser amplifier at 1, 2.5, 5 and 10 Gbit/s. Electronics Letters. 2002;**30**(4):338-339

[114] Arahira S, Ogawa Y. 160-Gb/s all-optical encoding experiments by four-wave mixing in a gain-clamped SOA with assist-light injection. Photonics Technology Letters, IEEE. 2004;**16**(2):653-655

[115] Li Z, Dong Y, Mo J, Wang Y, Lu C. Cascaded all-optical wavelength conversion for RZ-DPSK signal based on four-wave mixing in semiconductor optical amplifier. Photonics Technology Letters, IEEE. 2004;**16**(7):1685-1687

[116] Chan K, Chan CK, Chen LK, Tong F. Demonstration of 20-Gb/s all-optical XOR gate by four-wave mixing in semiconductor optical amplifier with RZ-DPSK modulated inputs. Photonics Technology Letters, IEEE. 2004;**16**(3):897-899

[117] Kamatani O, Kawanishi S. Prescaled timing extraction from 400 Gb/s optical signal using a phase lock loop based on four-wave-mixing in a laser diode amplifier. Photonics Technology Letters, IEEE. 2002;**8**(8):1094-1096

[118] Kamatani O, Katagiri Y, Kawanishi S. 100-Gbit/s optical TDM add/drop multiplexer based on photonic downconversion and four-wave mixing. In: Optical Fiber Communication Conference and Exhibit, 1998. OFC'98., Technical Digest. San Jose, CA: IEEE; 1998. pp. 112-113

[119] Simos H, Argyris A, Kanakidis D, Roditi E, Ikiades A, Syvridis D. Regenerative properties of wavelength converters based on FWM in a semiconductor optical amplifier. Photonics Technology Letters, IEEE. 2003;**15**(4):566-568

[120] Lacey J, Madden S, Summerfield M. Four-channel polarization-insensitive optically transparent wavelength converter. Photonics Technology Letters, IEEE. 2002;**9**(10):1355-1357

[121] Contestabile G, Presi M, Ciaramella E. Multiple wavelength conversion for WDM multicasting by FWM in an SOA. Photonics Technology Letters, IEEE. 2004;**16**(7):1775-1777

[122] Watanabe S, Takeda S, Ishikawa G, Ooi H, Nielsen J, Sonne C. Simultaneous Wavelength Conversion and Optical Phase Conjugation of 200 Gb/s (5× 40 Gb/s) WDM Signal Using a Highly Nonlinear Fiber Four-wave Mixer. Vol. 5. Edinburgh, UK: IET; 2002. pp. 1-4

[123] Watanabe S, Takeda S, Chikama T. Interband Wavelength Conversion of 320 Gb/s (32× 10 Gb/s) WDM Signal Using a Polarization-insensitive Fiber Four-wave Mixer. Vol. 3. Madrid, Spain: IEEE; 1998. pp. 83-87

[124] Clausen HCHMAT, Palushani E, Galili M, Hu H, Ji H, Xu J, et al. Ultra-high-speed optical signal processing of serial data signals. In: ICTON'12. Coventry, UK: University of Warwick; 2012

[125] Jianji Dong XZ, Xu J, Huang D, Songnian F, ShumZhang P, X. 40 Gb/s all-optical NRZ to RZ format conversion using single SOA assisted by optical bandpass filter. Optics Express. 2007;**15**(6):2907-2914

[126] Schmidt-Langhorst C, Ludwig R, Galili M, Huettl B, Futami F, Watanabe S, et al. 160 Gbit/s all-optical OOK to DPSK in-line format conversion. In: Presented at the 2006 Eur. Conf. Opt. Commun., Cannes, France, Paper Th 4.3.5. pp. 37-38

[127] Kuo YH, Rong H, Sih V, Xu S, Paniccia M, Cohen O. Demonstration of wavelength conversion at 40 Gb/s data rate in silicon waveguides. Optics Express. 2006;**14**(24):11721-11726

[128] Wang J, Sun J, Zhang X, Huang D, Fejer MM. All-optical format conversions using periodically poled lithium niobate waveguides. IEEE Journal of Quantum Electronics. 2009;**45**(2):195-205

[129] Contestabile G, Maruta A, Sekiguchi S, Morito K, Sugawara M, Kitayama K. All-optical wavelength multicasting in a QD-SOA. IEEE Journal of Quantum Electronics. 2011;**47**(4):541-547

[130] Zarris GH-S, Hugues-Salas E, Gonzalez NA, Weerasuriya R, Parmigiani F, Hillerkuss D, et al. Field experiments with a grooming switch for OTDM meshed networking. Journal of Lightwave Technology. 2010;**28**(4):316-327

[131] Akashi Y, Matsui SE, Isawa S, Matsushita A, Matsumoto A, Matsushima Y, et al. Demonstration of all-optical logic gate device using MQW-SOA and 10 Gbps XNOR operation. Physica Status Solidi (a). 2019;**216**(1):1800529

[132] Sheng Y, Chen X, Krolikowski W. Direct femtosecond laser writing of nonlinear photonic crystals. In: Advances in Optics: Reviews, Book Series. Vol. 2. Canberra, Australia: International Frequency Sensor Association (IFSA); 2018

[133] Alishahi F, Fallahpour A, Zou K, Cao Y, Kordts A, Karpov M, et al. Experimental generation and time multiplexing of data-carrying nyquist sinc shaped channels from a single microresonator-based Kerr frequency comb. In: Optical Fiber Communication Conference. San Diego, CA: Optical Society of America; 2019. p. W3I.2

[134] Kaur S, Goyal R. All-optical decoder/demultiplexer with enable using SOA based Mach-Zehnderinterferometers. In: 2019 6th International Conference on Signal Processing and Integrated Networks (SPIN). Noida, India: IEEE; 2019. pp. 780-784

[135] Choudhary A, Liu Y, Vu K, Ma P, Madden S, Marpaung D, et al. Narrowband gain in chalcogenide waveguides for low-power RF

delay lines. In: Fiber Lasers and Glass Photonics: Materials through Applications. Vol. 10683. Strasbourg, France: International Society for Optics and Photonics; 2018. p. 1068308

[136] Li S, Koch J, Pachnicke S. Optical signal processing in the discrete nonlinear frequency domain. In: 2018 Optical Fiber Communications Conference and Exposition (OFC). San Diego, CA: IEEE; 2018. pp. 1-3

[137] Huang T, Wu Y, Xie Y, Cheng Z. A slot-waveguide-based polarization beam splitter assisted by epsilon-near-zero material. Photonics and Nanostructures-Fundamentals and Applications. 2019;**33**:42-47

Chapter 5

Optical Fibers Profiling Using Interferometric and Digital Holographic Methods

Hamdy Wahba, Wael Ramadan and Ahmed El-Tawargy

Abstract

Optical fibers are extensively used in modern technology such as sensing, short-distance and long-distance telecommunications. This motivated many researchers to present different methods in order to characterize optical fibers and determine their refractive index profiles. In addition, variation of refractive index profiles of optical fibers suffering mechanical stresses or other external effects reflects good information about the internal structure of these fibers and how they are responding to these external effects. Optical interferometry and digital holography methods are accurate and effective tools used to achieve this task. In this chapter, we illustrate the application of different types of optical interference on conventional, polarizing maintaining and thick optical fibers. Also, we illustrate some mathematical interpretations (and recently automatic analyzing of interference patterns) that have been used to reconstruct the accurate refractive index profiles of optical fibers. Some experimental interferograms and refractive index profiles are demonstrated as well.

Keywords: optical fiber, optical interferometry, refractive index profile, interferogram analyses, digital holography

1. Introduction

1.1 Optical fibers

An optical fiber is an extended cylindrical optical waveguide. In its simplest form, it consists of a core having a certain refractive index n_c and is surrounded by a clad (sometimes called skin) of refractive index n_{cl} (or n_s). An optical fiber is used to guide light through its core, from one end to another, based on the principle of total internal reflection which mandates that n_c must be always higher than n_{cl}. Basically, optical fibers are made of highly pure silica glass doped with some impurities in order to increase n_c or decrease n_{cl} [1–3]. Recently, polymeric optical fibers got more attention as alternatives of some glass based optical fibers [1, 4].

Optical fibers are involved in many technological applications such as telecommunications, sensing [4, 5]; fiber lasers and fiber amplifiers [6]; fiber gratings which can act as mirrors [7, 8]; mode converters [9]; modulators; and couplers and switches [10, 11]. Optical fibers are considered ideal optical transmission media since communication cables hundreds of kilometers in length can be obtained with

 IntechOpen

low absorption and low loss due to the purity and cross-sectional uniformity of the manufactured optical fibers. Moreover, accurate tuning of the refractive indices of both core and clad guarantee extremely low scattering loss at the interfaces [1].

1.2 Types of optical fibers

The commonly known optical fibers are step index and graded index (GR-IN) optical fibers. The former means that the core's refractive index is homogeneous while it suffers an abrupt change at the boundary with the clad. For a GR-IN optical fiber, the core does not have a constant value of refractive index but it rather has a radial distribution of refractive index. These two types of optical fibers can be classified into either single-mode or multi-mode optical fibers. Single-mode optical fiber only sustains one mode of propagation while the multi-mode optical fiber can sustain up to hundreds of propagation modes [1, 3]. The number of the propagation modes is related to the numerical aperture of the fiber, which, in turn, depends on the refractive indices of both core and clad.

1.3 Characterization of optical fibers

Accurate characterization of optical fibers is required in order to know about their functions and performances. There are many methods of optical fibers characterization such as optical microscopy, electron microscopy, X-ray spectrometry, infrared spectroscopy, light diffraction, light scattering, optical interferometry, and digital holography [1, 3, 12–29]. Optical interferometry is an effective accurate tool for studying and characterizing optical fibers. It depends on the determination of the phase difference between a ray of light transmitting the fiber's cross-section and a reference ray reaching the interference plane directly without crossing the fiber. This phase shift can be transformed into a refractive index map representing the radial distribution of the refractive index across the fiber or, in other words, the refractive index profile (RIP). Interferometry can detect tiny changes in refractive index if an external effect is applied on the fiber. The change of refractive index can be in situ detected if the interferometer is developed to achieve this task. Interference patterns can be digitally processed and analyzed in order to increase the accuracy of the obtained results [1, 17, 22, 26, 27, 30–35].

Interference techniques can be classified into either two-beam interferometers such as Michelson, Mach-Zehnder, Pluta polarizing microscope, Lioyd's mirror, etc., or multiple-beam interferometers such as Fabry-Pérot and Fizeau interferometers [1, 3, 25, 36–39]. A two-beam interferometer produces a pattern of alternate bright and dark fringes of equal thicknesses when two beams, usually, of equal intensities I_o suffering a relative phase difference δ are superposed. The resultant intensity distribution I of the interference pattern is given as:

$$I = I_o \cos^2\left(\frac{\delta}{2}\right) \tag{1}$$

Multiple-beam interference takes place when light rays fall on two parallel optical plates enclosing a small distance between each other while their inner surfaces are highly reflecting and partially transparent. The intensities of both reflected, $I^{(r)}$, and transmitted, $I^{(t)}$, light distributions that are redistributed due to the multiple-beam interference are given as [40]

$$I^{(r)} = \frac{(2 - 2\cos\delta)R}{1 + R^2 - 2R\ \cos\ \delta} I^{(i)} \tag{2}$$

$$I^{(t)} = \frac{T^2}{1 + R^2 - 2R\ \cos\ \delta} I^{(i)} \tag{3}$$

where, $I^{(i)}$ is the intensity of the incident light, T and R are the products of the transmission and reflection coefficients of the two surfaces, respectively, while δ is the phase difference between any two consecutive interfered rays.

On the other hand, holography was firstly presented by Gabor in 1947 as a lens-less process for image formation by reconstruction of wave-fronts [41–43]. It offers 3D characterization such as the depth of field from recording and reconstructing the whole optical wave field, intensity, and phase [41, 42]. Holographic interferometry is a non-destructive, contactless tool that can be used for measuring shapes, deformations and refractive index distributions [44, 45]. The modern digital holography was introduced in 1994 [46–48]. Moreover, the phase shifting interferometric (PSI) technique was introduced by Hariharan et al. as an accurate method for measuring interference fringes in the real time [49]. Recently, digital holographic phase shifting interferometry (DHPSI) was used to investigate some optical parameters of fibrous materials [17, 18, 21, 26–29].

1.4 Digital holographic phase shifting interferometry (DHPSI)

In DHPSI, frequently a set of four [20] or five [23, 33] phase shifted holograms with known mutual phase shifts starting with 0^0 and having 90^0 separations have to be recorded [21]. These recorded holograms can be represented by:

$$I_n = a(\zeta,\eta) + b(\zeta,\eta)\cos\left(\varphi(\zeta,\eta) + \varphi_{Rn}\right), n = 1, 2, 3, \ldots \tag{4}$$

where $a(\zeta,\ \eta)$ and $b(\zeta,\ \eta)$ are the additive and the multiplicative distortions and φ_{Rn} is the phase shift of the reference wave. In case of four and five phase shifted holograms, the complex wavefield [37] in the hologram plane can be calculated using Eqs. (5) and (6), respectively.

$$h(\zeta,\eta) = [I_1(\zeta,\eta) - I_3(\zeta,\eta)] + i[I_4(\zeta,\eta) - I_2(\zeta,\eta)] \tag{5}$$

or,

$$h(\zeta,\eta) = [4I_1(\zeta,\eta) - I_2(\zeta,\eta) - 6I_3(\zeta,\eta) - I_4(\zeta,\eta) + 4I_5(\zeta,\eta)] + i[7(I_4(\zeta,\eta) - I_2(\zeta,\eta))] \tag{6}$$

In digital holography, the recorded wavefield is reconstructed, based on Fresnel diffraction integral, by multiplying the stored hologram by the complex conjugate of the reference wave $r^*(\zeta,\ \eta)$ to calculate the diffraction field $b'(x', y')$ in the image plane, see **Figure 1**. This can be calculated using the finite discrete form of the Fresnel approximation to the diffraction integral as:

$$b'(n\Delta x', m\Delta y') = A\sum_{j=0}^{N-1}\sum_{l=0}^{M-1} h(j\Delta\zeta, l\Delta\eta)r^*(j\Delta\zeta, l\Delta\eta) \times \exp\left\{\frac{i\pi}{d'\lambda}\left(j^2\Delta\zeta^2 + l^2\Delta\eta^2\right)\right\}\exp\left\{2i\pi\left(\frac{jn}{N} + \frac{lm}{M}\right)\right\} \tag{7}$$

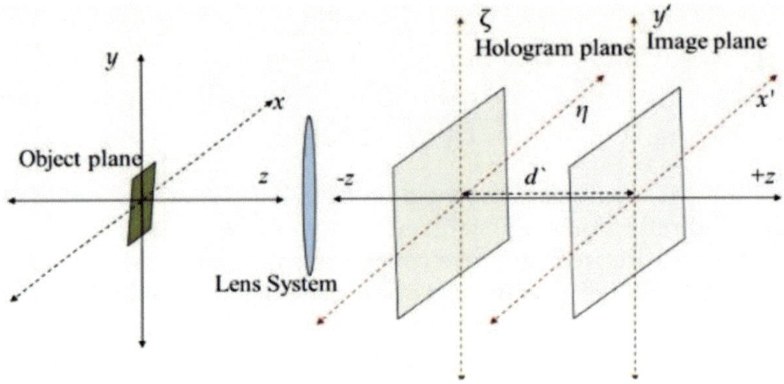

Figure 1.
Geometry of digital holographic axes and the planes systems.

The parameters used in this formula depend on the used CCD array of $N \times M$ pixels and the pixel pitches $\Delta\zeta$ and $\Delta\eta$. The distance between the hologram and the image plane is denoted by *d'*. The pixel spacings in the reconstructed field of image are:

$$\Delta x' = \frac{d'\lambda}{N\Delta\zeta} \; and \; \Delta y' = \frac{d'\lambda}{M\Delta\eta} \tag{8}$$

The convolution of $h(\zeta, \eta)r^*(\zeta, \eta)$ can be used as alternative of Fresnel approximation [37]. The resulting pixel spacing for this convolution approach is

$$\Delta x' = \Delta\zeta \; and \; \Delta y' = \Delta\eta \tag{9}$$

In addition, the phase shifted holograms are used to overcome the problems of the d.c. term and twin image, in which the calculated complex wavefield is used instead of a real hologram in the convolution approach.

The intensity and phase distributions in the reconstruction plane are given by

$$I(x', y') = |b'(x', y')|^2 \tag{10}$$

$$\varphi(x', y') = arctan\left\{\frac{Im|b'(x', y')|}{Re\,|b'(x', y')|}\right\} \tag{11}$$

So, the optical phase differences due to phase objects can be extracted.

Mach-Zehnder interference-like system is used as a digital holographic setup as shown in **Figure 2** [20, 23, 29, 33]. The optical waveguide sample, such as optical fiber, is immersed in a liquid of refractive index n_L near or matching the cladding refractive index n_{clad} of the sample. The interference patterns are recorded using a charge-coupled device, that is, CCD camera.

In this chapter, we illustrate some featured work on interferometric characterization (sometimes, implying digital holographic interferometry) of different optical fibers done by our research group during the last three decades. In Section 2, interferometric characterization of conventional step-index and GR-IN optical fibers is presented. Section 3 illustrates characterization of the conventional optical fibers when they are suffering mechanical bending. In Section 4, interferometric characterization of a special type of optical fibers called polarization maintaining (PM) optical fibers is presented. In the last section, we elucidate thick optical

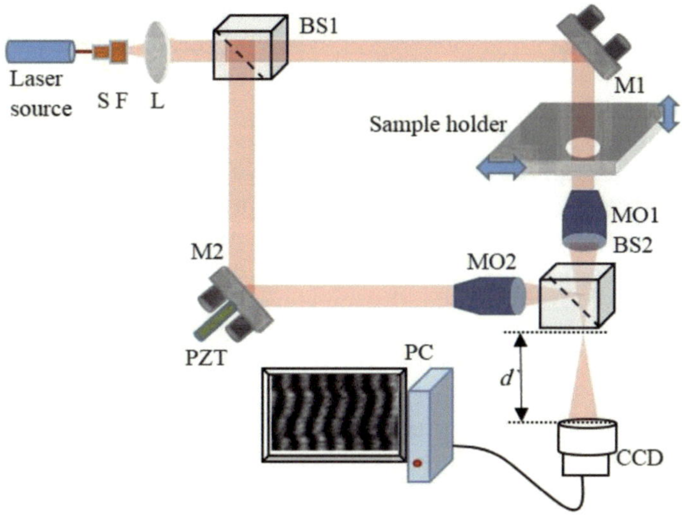

Figure 2.
Mach-Zehnder digital holographic interferometric set-up, S F: spatial filter, L: collimating lens, BS: beam splitter, M: mirror, and MO: microscopic objective.

fibers and their interferometric characterization with a special interferometric system, developed in our laboratory, called lens-fiber interferometry (LFI).

2. Conventional optical fibers

2.1 Step-index optical fiber

In 1994, Hamza et al. derived a mathematical expression to calculate the RIP of an optical fiber by considering the refraction of optical rays at the liquid-clad and clad-core interfaces, see **Figure 3** [12]. It was the first time to consider the refraction of the transmitted rays to reconstruct the RIP of a fiber. The derived expressions for calculating the RIP in case of two-beam and multiple-beam interferences, based on **Figure 3**, are given by Eqs. (12) and (13), respectively.

$$\frac{Z(x)\lambda}{h} = 2n_s\left\{\sqrt{\left[R^2-\left(\frac{n_L d}{n_s}\right)^2\right]}-\sqrt{\left[(R-e)^2-\left(\frac{n_L d}{n_s}\right)^2\right]}\right\} + 2n_c\sqrt{\left[(R-e)^2-\left(\frac{n_L d}{n_c}\right)^2\right]} - n_L\left[\sqrt{(R^2-d^2)}+\sqrt{(R^2-x^2)}\right] \quad 0 \le d < R \tag{12}$$

$$\frac{Z(x)\lambda}{2h} = 2n_s\left\{\sqrt{\left[R^2-\left(\frac{n_L d}{n_s}\right)^2\right]}-\sqrt{\left[(R-e)^2-\left(\frac{n_L d}{n_s}\right)^2\right]}\right\} + 2n_c\sqrt{\left[(R-e)^2-\left(\frac{n_L d}{n_c}\right)^2\right]} - n_L\left[\sqrt{(R^2-d^2)}+\sqrt{(R^2-x^2)}\right] \quad 0 \le d < R \tag{13}$$

where, R is the fiber's radius and e is the skin's thickness. n_L, n_s, and n_c are the refractive indices of the immersion liquid, skin, and core, respectively. λ is the wavelength of the used illuminating source. L_s and L_c are the geometrical path lengths inside the skin and the core, respectively. Z is the fringe shift due to the presence of the fiber while h is the interfringe spacing and d is the distance measured from the center of the fiber to the position of the incident ray.

In that work, they used Fizeau interferometer to determine the refractive index profile of FOS Ge-doped step-index multi-mode optical fiber with a core radius 19.5 μm. The fiber was immersed in a liquid of refractive index n_L = 1.4665, which was a little bit greater than n_s while the wavelength of the used illuminating source was λ = 546.1 nm. The Fizeau interferogram of this fiber is shown in **Figure 4a**. The obtained RIP was compared with the profile calculated for the same fiber when the refraction of light through the fiber was neglected as was usually done by other authors before this work. There was a significant difference between the two profiles, see **Figure 4b**. Therefore, the refraction through the fiber was recommended to be considered for calculating RIPs particularly when the refractive index of the immersion liquid is not close to the fiber's refractive index.

In 2008, another mathematical model was derived in order to determine RIPs of fibers having regular and/or irregular cross-sections [38]. This method was based

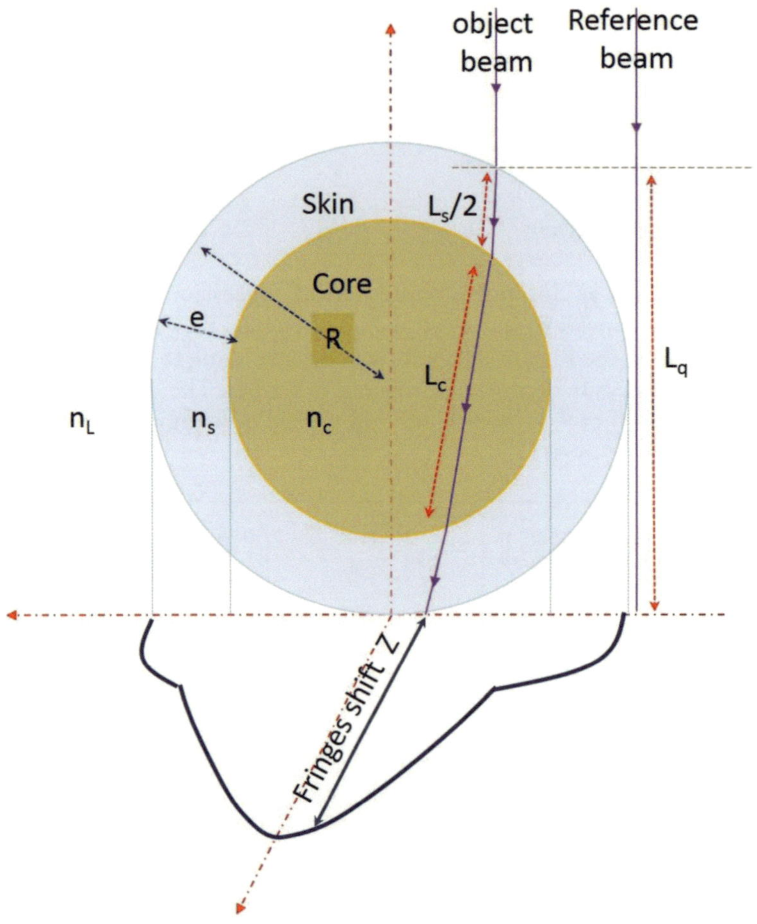

Figure 3.
An incident ray (object ray) is refracted due to a clad-core fiber causing a fringe shift Z when interferes with a reference ray.

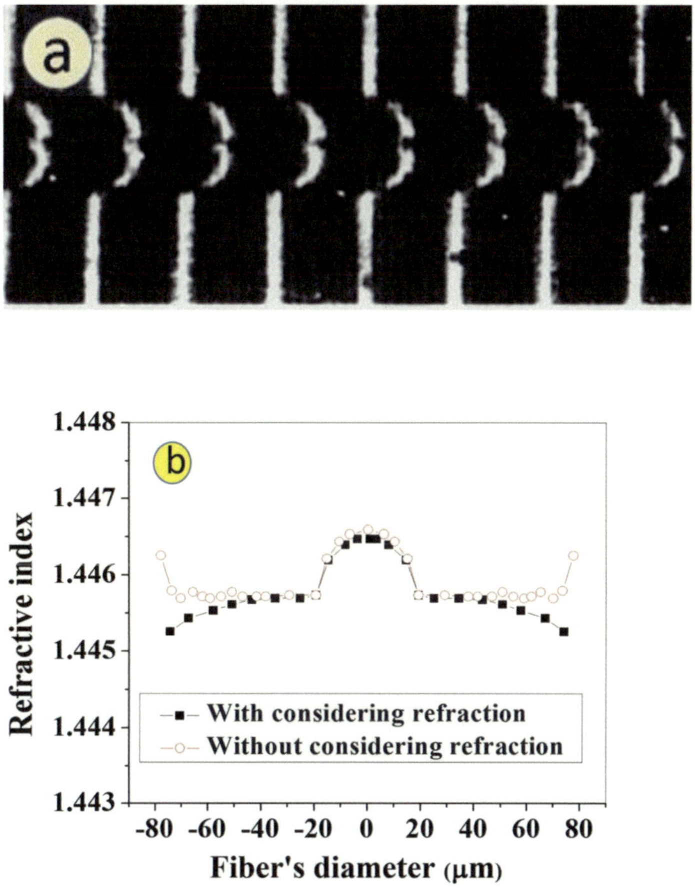

Figure 4.
(a) An interference pattern of Fizeau fringes, in transmission, for a FOS step-index optical fiber. (b) RIPs of this fiber in case of considering and neglecting the retraction of the crossing rays inside the fiber.

on immersing the investigated fiber in two liquids with different, but so closed, refractive indices. They applied this method on a single-mode optical fiber, having a small core of radius <5 μm while the fiber's radius was 60.6 μm, as shown by Fizeau interferograms in **Figure 5** when the fiber was immersed in two liquids with refractive indices (a) 1.4589 and (b) 1.4574. The obtained RIP of this fiber is illustrated in **Figure 5c** showing that this fiber has n_c = 1.4630 and n_{cl} = 1.4596. This method was simple and accurate enough to detect such a small core of a step-index optical fiber.

2.2 Graded-index (GR-IN) optical fiber

A GR-IN optical fiber with a radial refractive index distribution was suggested to be divided into a finite number (M) of concentric layers where each layer has its own value of refractive index, see **Figure 6a**. The thickness (a) of each layer equals R/M, where R is the radius of the graded-index part. When the ray falls on the fiber at a distance d_Q apart from the fiber's center, the ray refracts through Q layers. The nearest layer to the fiber's center has a refractive index n_Q. The fiber's RIP can be calculated using Eq. (14) in case of two-beam interference and Eq. (15) in case of multiple-beam interference [13]. Another model was presented in order to get RIP of a GR-IN optical fiber by considering the real path of the optical ray due to the

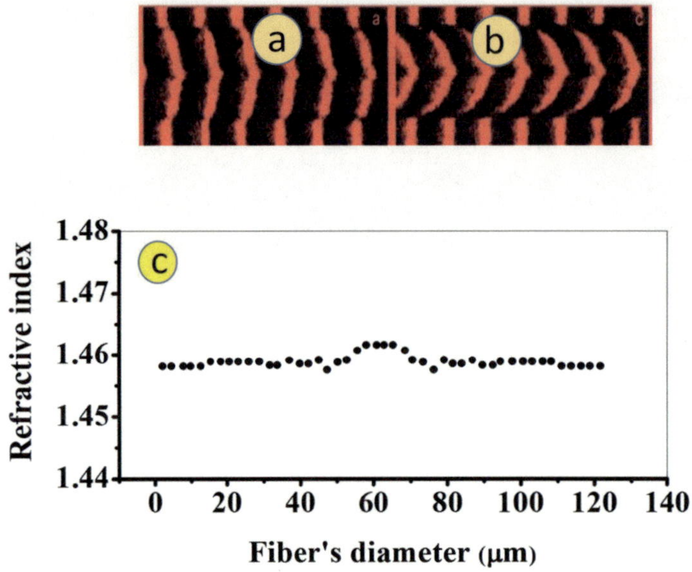

Figure 5.
Fizeau interferograms, in transmission, for a single-mode optical fiber when it was immersed in two liquids of refractive indices (a) 1.4589, (b) 1.4574. (c) RIP of the single-mode optical fiber having the interferograms shown in (a) and (b).

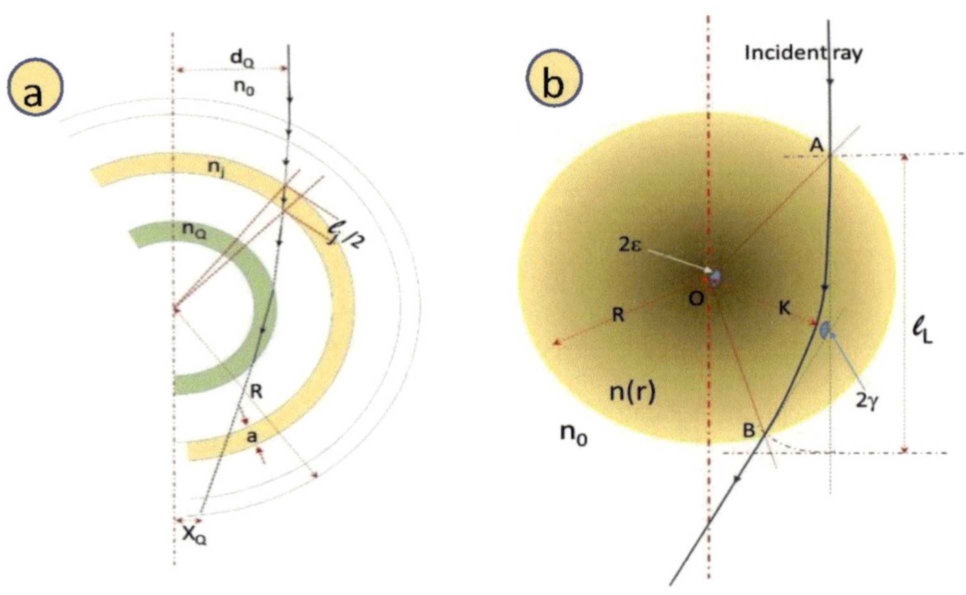

Figure 6.
(a) A schematic diagram shows the path of an optical ray crossing Q layers in the core region. (b) A schematic diagram shows the path of an optical ray crossing a GR-IN core optical fiber.

refraction in the core region as well as adding a correction for the ray passing through the immersion liquid [50], see **Figure 6b**. In this case, the fringe shift was obtained by assuming values for both the profile shape parameter (α) and the difference between refractive indices of core and clad (Δn). A prepared software was programmed to iterate and get the best values of α and Δn and comapre the calculated fringe shift with the experimentally obtained one.

$$\frac{\lambda Z_Q}{h} = \sum_{j=1}^{Q-1} 2n_j \left[\sqrt{(R-(j-1)a)^2 - \left(\frac{d_Q n_o}{n_j}\right)^2} - \sqrt{(R-ja)^2 - \left(\frac{d_Q n_o}{n_j}\right)^2} \right] + 2n_Q \sqrt{(R-(Q-1)a)^2 - \left(\frac{d_Q n_o}{n_Q}\right)^2} - n_o \left[\sqrt{R^2 - d_Q^2} + \sqrt{R^2 - x_Q^2} \right] \quad (14)$$

$$\frac{\lambda Z_Q}{2h} = \sum_{j=1}^{Q-1} 2n_j \left[\sqrt{(R-(j-1)a)^2 - \left(\frac{d_Q n_o}{n_j}\right)^2} - \sqrt{(R-ja)^2 - \left(\frac{d_Q n_o}{n_j}\right)^2} \right] + 2n_Q \sqrt{(R-(Q-1)a)^2 - \left(\frac{d_Q n_o}{n_Q}\right)^2} - n_o \left[\sqrt{R^2 - d_Q^2} + \sqrt{R^2 - x_Q^2} \right] \quad (15)$$

According to **Figure 6b**, the optical pathlengths of the ray crossing the core $Ol(K)$ and the ray passing only in the immersing liquid Ol_L are given by the following relations.

$$Ol(K) = 2\int_K^R \frac{n^2(r)r}{\sqrt{n^2(r)r^2 - n^2(K)K^2}} dr \quad (16)$$

$$Ol_L = 2n_o R \frac{\sin(\varepsilon)}{\sin(\gamma)} \quad (17)$$

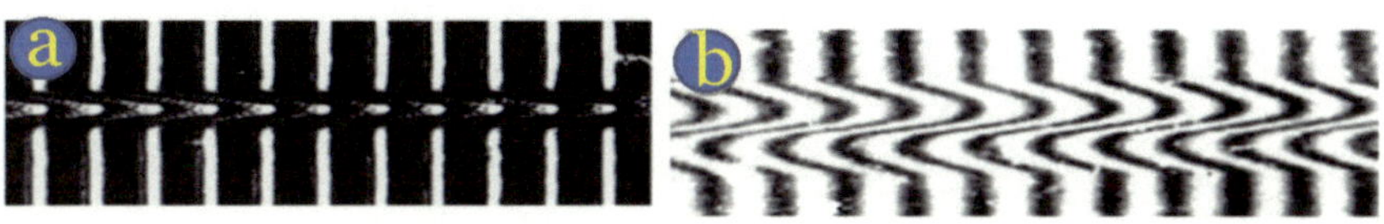

Figure 7.
(a) Pluta duplicated image of LDF GR-IN optical fiber and (b) Fizeau interferogram of the same sample. Reference [50] with permission.

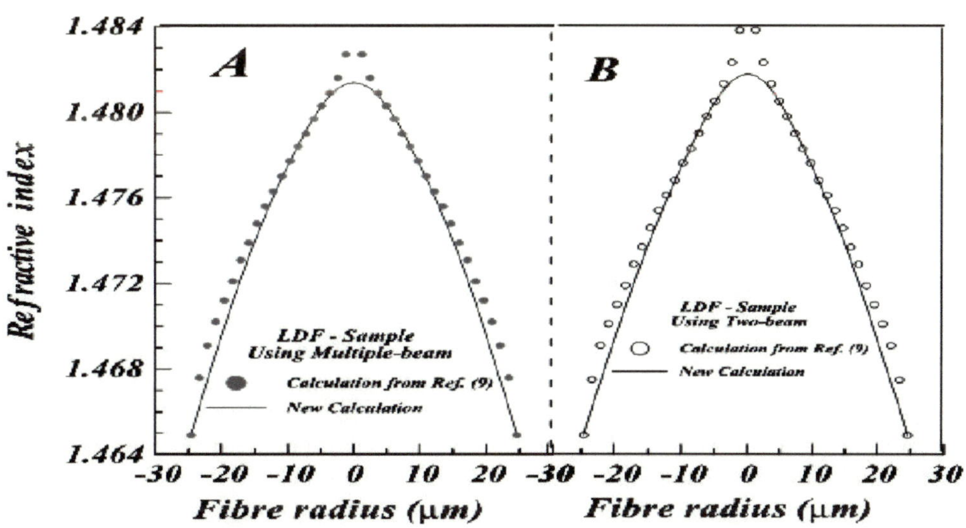

Figure 8.
A comparison between RIPs of LDF GR-IN optical fiber using the model in Ref. [27] (dots) and model in Ref. [28] (solid curve) in case of (a) multiple-beam Fizeau interference and (b) two-beam Pluta interference. Ref. [50] with permission.

where, R is the core's radius, k is the minimum distance between the fiber's center and the bent ray, ε is the half of the angle determined by the two radii that are enclosing the bent ray inside the graded-index region, and γ is the half of the angle between the incident and the emerged rays. **Figure 7** shows the interferograms of LDF GR-IN optical fiber when it was investigated by (a) Pluta and (b) Fizeau interferometers. **Figure 8** shows the RIPs calculated by these last models for the LDF optical fiber. The last model, presented in 2001 [50], provided more accurate values of the RIP of a GR-IN optical fiber compared with its previous presented model in Ref. [13].

However, the former requires knowing the function describing the index profile while the aim is to find the parameters of this function.

3. Bent conventional optical fibers

Optical fibers, which are isotropic materials, can suffer a birefringence under external mechanical bending effects [1, 22, 33, 51]. The induced birefringence can be used in sensing applications [52–54]. However, bending has an unfavorable effect on the optical fibers used in telecommunications where it, sometimes, causes a mode disturbance and consequently a signal attenuation [55, 56]. An approach to calculate the refractive index profile of a bent optical fiber was proposed where the fiber was divided into layers and slabs simultaneously [22]. The refraction of the optical rays at the liquid-clad and clad-core interfaces was considered. Unfortunately, this approach did not consider the change of refractive index inside each slab. Also, the expected change of refractive index due to the release of stresses near the fiber's free surface has not been considered. However, this approach succeeded to present good information about the variation of mode propagation due to bending.

3.1 Step-index bent conventional optical fiber

In 2014, Ramadan et al. calculated the refractive index and the induced birefringence profiles of bent step-index optical fibers using digital holographic Mach-Zehnder interferometer [33]. In that work, they considered two different processes controlling the variations of the refractive index of the bent fiber: (1) the linear refractive index variation due to the applied stress along the bent radius and (2) the release of this stress on the fiber's surface. The first one is dominant when approaching the center of the fiber while the second one is dominant near the fiber's free surface and decays on moving toward the fiber's center. **Figure 9** shows the difference between the paths of optical rays through the bent fiber in the compressed and expanded parts. The stress release was supposed to have a radial dependence on the fiber's radius, which enabled the construction of 2D RIP of the investigated bent homogeneous optical fiber. Based on the expected stress values due to the bending effect, a function describing the RIP was proposed and used to integrate the optical path of the ray traversing the fiber [50]. By adapting the appropriate parameters of this function, the optical phase differences were estimated and matched those phase differences that were experimentally obtained. By this assumption, a realistic induced stress profile due to bending was obtained [33]. DHPSI was used in that study where the recorded phase shifted holograms were combined and processed to extract the phase map of the fiber [18]. By considering both of the mentioned effects, the following function was chosen to describe the RIP of the bent optical fiber [33].

$$n(x,r) = -\rho n_{bf}\frac{x}{R}\left[1 - e^{-\left(\frac{r_Q - r}{r_s}\right)}\right] + n_{cl} \tag{18}$$

where ρ is the strain-optic coefficient, n_{bf} is the refractive index of the bent-free fiber, R is the radius of bending, r_o is the radius of the fiber, n_{cl} is the clad's refractive index, r_s is the proposed parameter to control the distance suffering stress release from the surface of the fiber, and x is the distance between the center of the fiber and the position of the incident ray.

The first term of Eq. (18) gives the bent-induced birefringence,

$$\Delta n(x,r) = -\rho n_{bf}\frac{x}{R}\left[1 - e^{-\left(\frac{r_Q - r}{r_s}\right)}\right] \tag{19}$$

which is correlated to the generated stress $S\ (r,x)$ inside the fiber

$$S(r,x) = E\left(\frac{-\Delta n(x,r)}{\rho n_{bf}}\right) = E\frac{x}{R}\left[1 - e^{-\left(\frac{r_Q - r}{r_s}\right)}\right] \tag{20}$$

Eq. (20) evaluates the distribution of stress over the fiber's cross-section for different bending radii where E is the Young's modulus of the bent fiber. The signs of Δn are opposite to the signs of tensile and compressive stresses. The tensile stress was chosen to be positive.

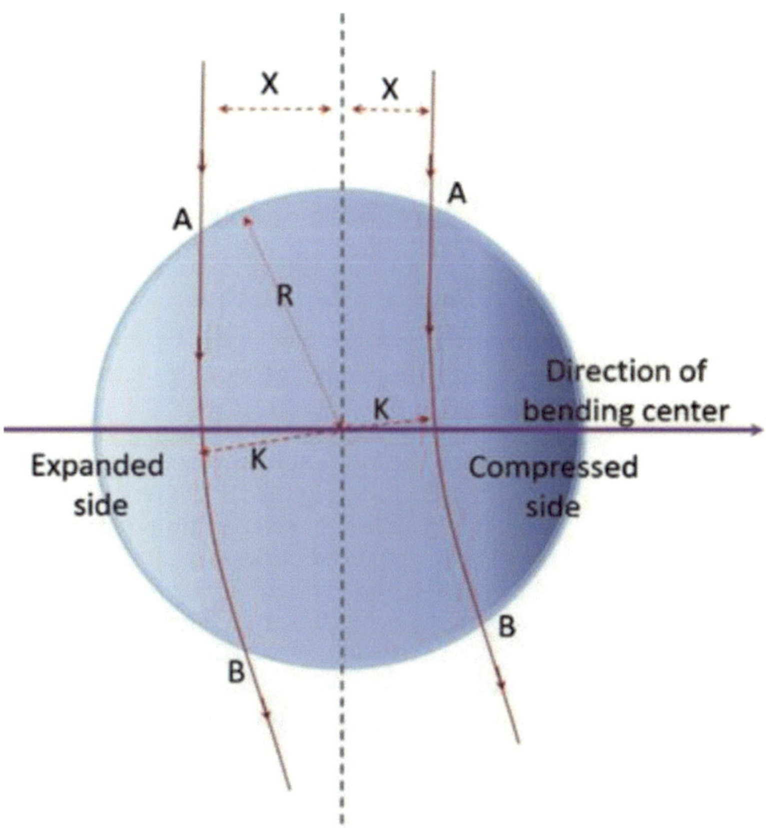

Figure 9.
A schematic diagram shows the path of an optical ray crossing a bent homogeneous optical fiber.

Since bending such a step-index optical fiber converts it into a weekly graded-index fiber, Bouguer's formula [40] was used to correlate the radius, incidence angles, and refractive index of the bent fiber as follows:

$$Kn(x,K) = rn(x,r)\sin\theta_r \tag{21}$$

where $n(x,r)$ is the refractive index at radius r. By applying this formula at the incidence point, one obtains

$$Kn(x,K) = r_o n_L \frac{x}{r_o} = xn_L \tag{22}$$

This equation was numerically solved to get K satisfying the lower integration limit of the optical path difference for a certain value of x. Based on the model described in Ref. [50], the infinitesimal change in the geometrical distance along the path of the optical ray with respect to the radius variation was given as:

$$\left(\frac{\partial l}{\partial r}\right)_{n(x,r)} = \frac{2n(x,r)r}{\sqrt{n^2(x,r)r^2 - n^2(x,k)k^2}} \tag{23}$$

By integration with respect to r, the total path length inside the fiber is:

$$l(x) = 2\int_K^{r_o} \frac{n(x,r)r}{\sqrt{n^2(x.r)r^2 - (n_L x)^2}} dr \tag{24}$$

The optical path length difference between this ray, passed through the fiber, and the reference ray passed through the liquid is:

$$opld(x) = 2\int_K^{r_o} \frac{(n(x,r) - n_L)\cdot n(x,r)r}{\sqrt{n^2(x.r)r^2 - (n_L x)^2}} dr \tag{25}$$

The phase difference is given as:

$$\phi(x) = \frac{2\pi}{\lambda} opld(x) \tag{26}$$

Figure 10 shows a set of five shifted holograms of a bent step-index optical fiber with a bending radius R = 8 mm when the incident light was vibrating parallel to the fiber's axis. They were recorded in order to apply the DHPSI technique and reconstruct the RIP of the bent fiber. The 2π shifted interferogram was analyzed and its reconstructed interference phase map, enhanced phase map, and interference phase distribution are shown in **Figures 11a–c**, respectively. The refractive index cross-section distribution of the bent optical fiber is shown in **Figure 12** while the strain-optic coefficients in compression and expansion were 0.208 and 0.224, respectively.

3.2 Graded-index bent conventional optical fiber

In 2017, Ramadan et al. presented a theory to recover the RIP of a bent GR-IN optical fiber inside the core region using DHPSI [35]. They assumed the two different processes controlling the shape of the RIP: (1) the linear variation due to stresses in the direction of the bent radius and (2) the release of the stresses near the fiber's surface.

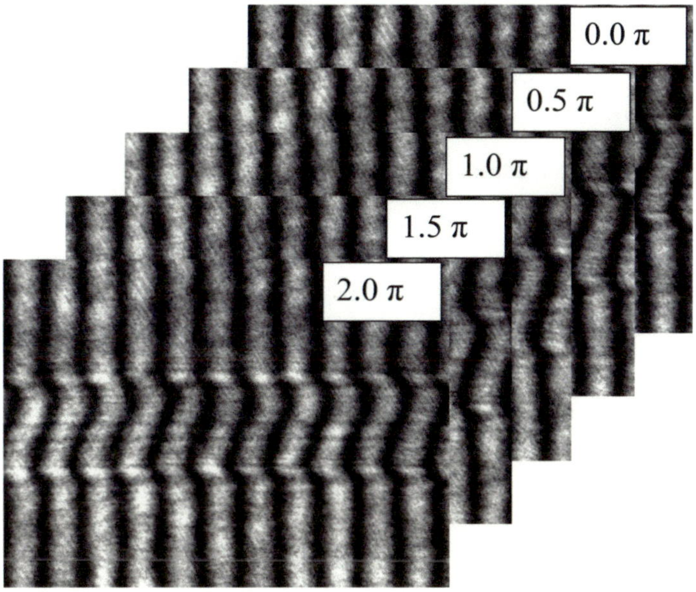

Figure 10.
A set of five shifted interferograms of a bent step-index optical fiber.

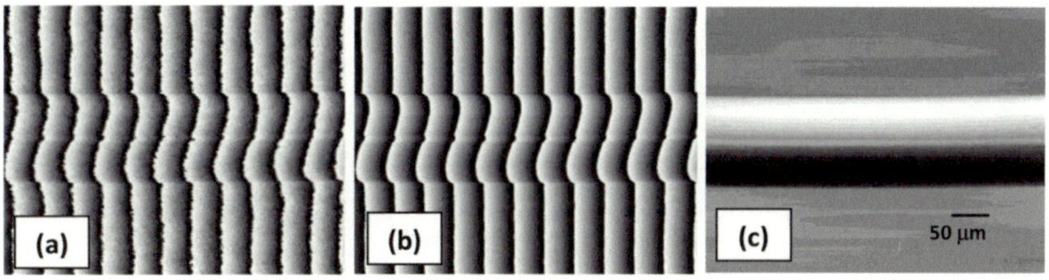

Figure 11.
(a) The reconstructed interference phase map modulo 2π, (b) its enhanced phase map, and (c) the interference phase distribution.

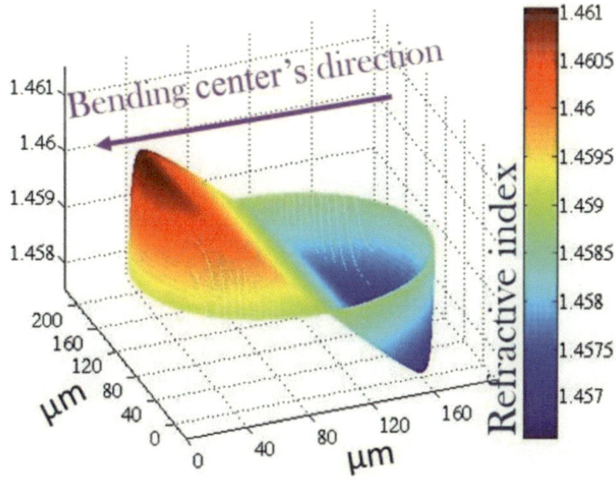

Figure 12.
The refractive index cross-section distribution of the bent optical fiber, R = *8.*

The total optical path length of the optical ray crossing the bent GR-IN optical fiber is given by Eq. (27), see **Figure 13**. The calculated optical path length differences of the interfered rays can be transformed, afterward, into a phase difference map using Eq. (26).

$$OPlD(d) = OPlD_{cl}(d) + OPlD_c(d) \tag{27}$$

with,

$$OPlD_{cl}(d) = 2\int_{r_c}^{r_{cl}} \frac{(n_{cl}(d,r) - n_L)n_d(d,r)r}{\sqrt{n_{cl}(d,r)^2 r^2 - n_{cl}(d,k_{cl})^2 k_{cl}^2}} dr \tag{28}$$

$$OPlD_c(d) = 2\int_{k_c}^{r_c} \frac{(n_c(d,r) - n_L)n_c(d,r)r}{\sqrt{n_c(d,r)^2 r^2 - n_c(d,k_c)^2 k_c^2}} dr \tag{29}$$

Figure 14a shows a set of five phase shifted interferograms for the bent GR-IN optical fiber with bending radius R = 8 mm when the incident light was vibrating parallel to the fier's axis. The enhanced reconstructed phase modulo 2π and the interference phase distribution of the bent fiber are shown in **Figure 14b**. Due to

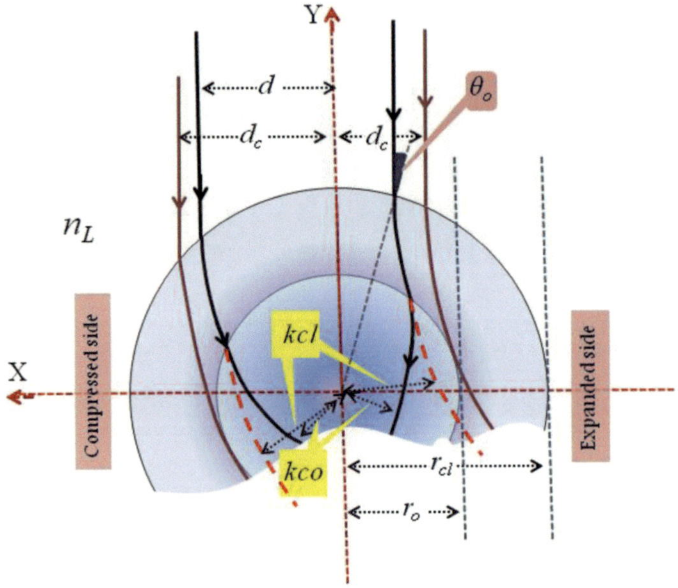

Figure 13.
schematic diagram shows the ray tracing in case of traversing bent GR-IN fiber.

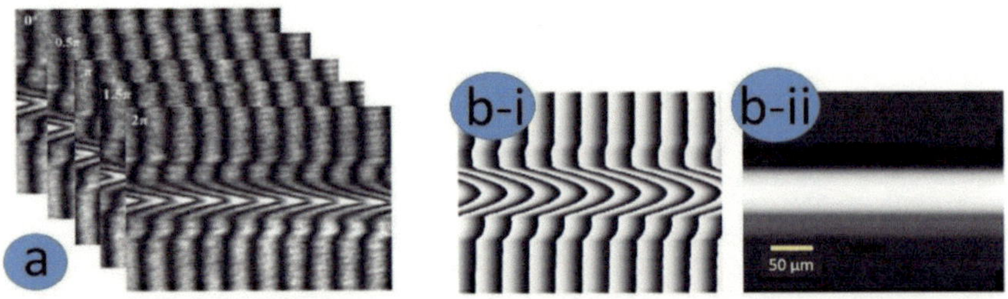

Figure 14.
(a) A set of five phase shifted interferograms of a bent GR-IN optical fiber. (b) The enhanced reconstructed phase modulo 2π and the interference phase distribution. Ref. [35] with permission.

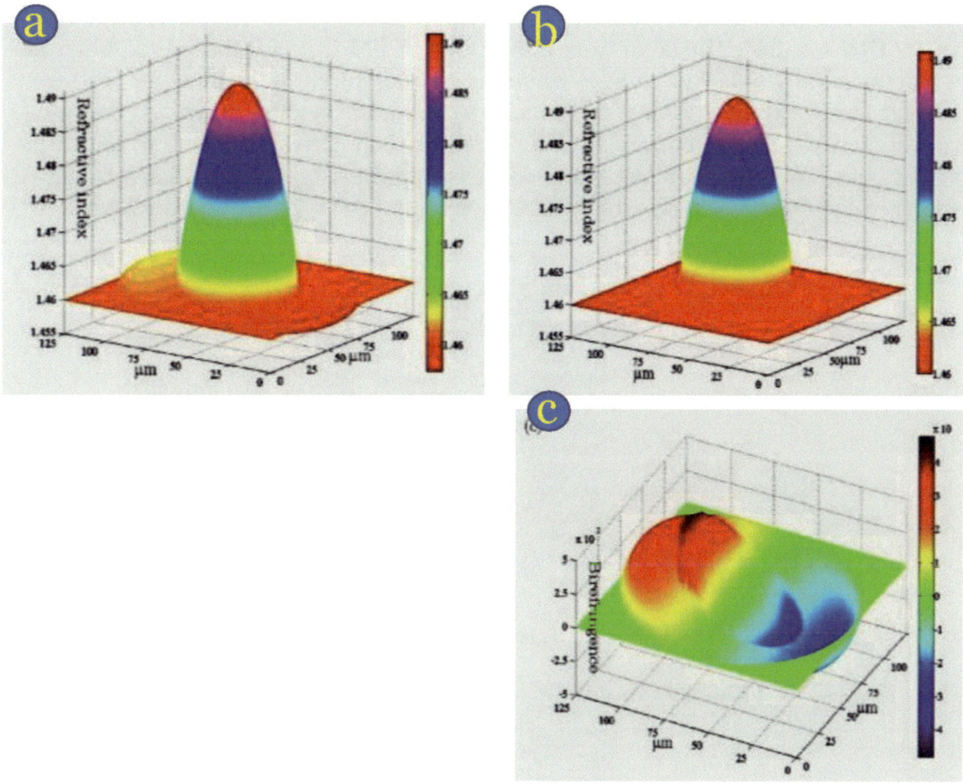

Figure 15.
Refractive index cross-section distribution of the bent GR-IN optical fiber when the incident light vibrates (a) parallel and (b) perpendicular to the fiber's axis. (c) The birefringence cross-section distribution, R = *8 mm. Ref. [35] with permission.*

the bending process, the GR-IN optical fiber exhibited a birefringence where the RIPs when the incident light vibrated parallel and perpendicular to the fiber's axis were different, see **Figure 15**.

4. Polarization maintaining (PM) optical fibers

A PM fiber is any fiber that preserves and transmits the polarization state of the light launched into the fiber even if this fiber is subjected to environmental

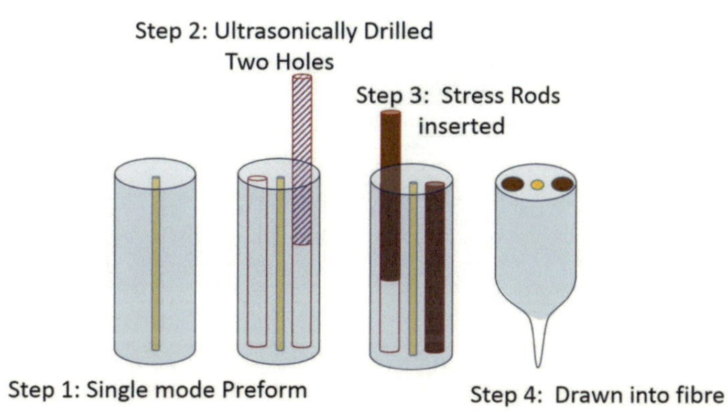

Figure 16.
Manufacturing steps of a PANDA PM optical fiber.

perturbations [57]. This advantage cannot be verified by conventional single-mode optical fibers outside the laboratory conditions. A PM fiber is tailored to oblige the two orthogonally polarized modes traveling with different velocities (i.e., different propagation constants). This difference in velocities prevents the optical energy from suffering a "cross-coupling" and preserves the polarization state of the transmitted light. Therefore, a PM fiber used in any application requires delivering a polarized light such as in telecommunications, medical applications, and sensing.

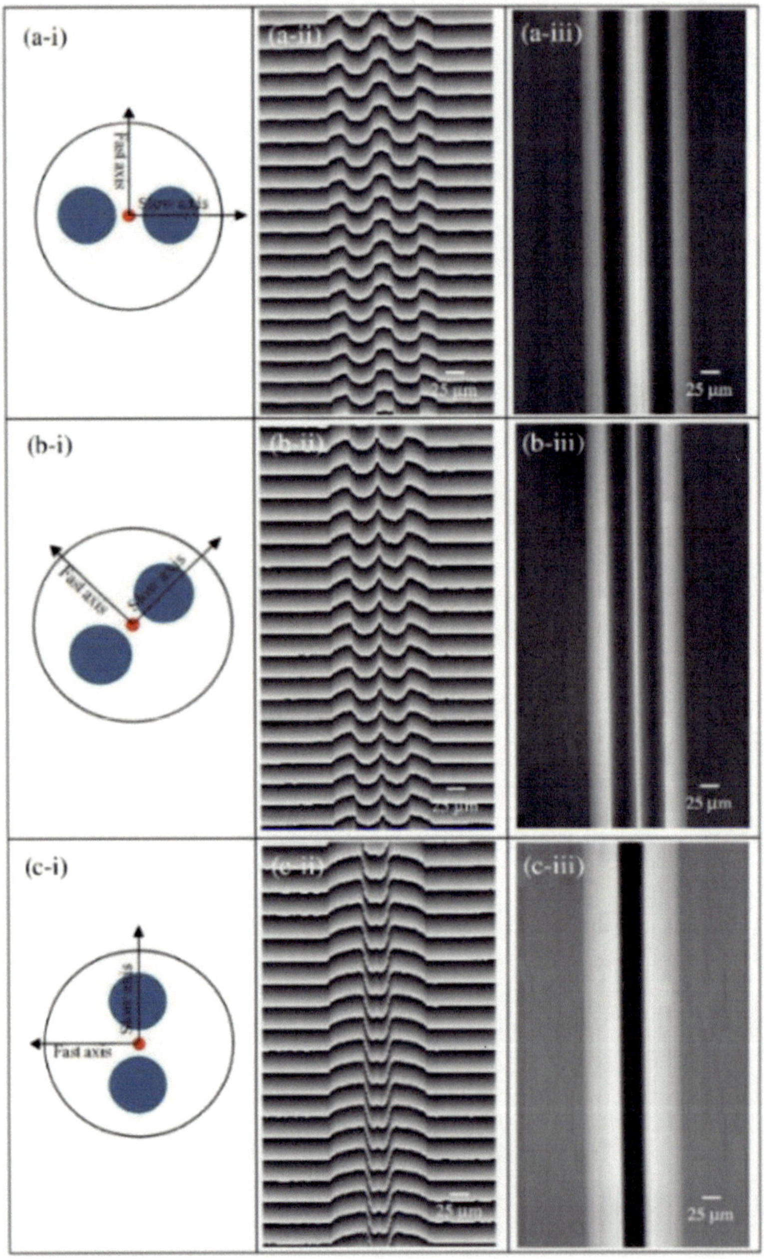

Figure 17.
The left column shows three orientations of PANDA PM optical fiber as it was rotated during the characterization process where the slow axis makes an angle (a-i) 0°, (b-i) 45° and (c-i) 90° with the horizontal axis. The middle column shows their reconstructed interference phase modulo 2π while the right column shows their phase difference maps. Ref. [23] with permission.

In interferometric applications, it is used to affirm that the interfered rays have the same polarization states. To maintain such a difference of velocities, the core of the fiber has to be anisotropic either geometrically by making the core cross-section as an ellipse or by applying a uniaxial stress. The most known PM fibers used today are, PANDA, bow tie, and elliptical-jacket fibers. These types are designed by the same way where the cores are flanked by areas of high-expansion glass and shrunk-back more than the surrounding silica then the core is frozen under tension. The birefringence is induced due to this tension, which means creation of two different indices of refraction: a higher index in the direction parallel to the applied stress and a lower index perpendicular to the direction of the applied stress. In the next two subsections, we briefly illustrate both the manufacturing process and interferometric characterization of PANDA and bow tie PM optical fibers.

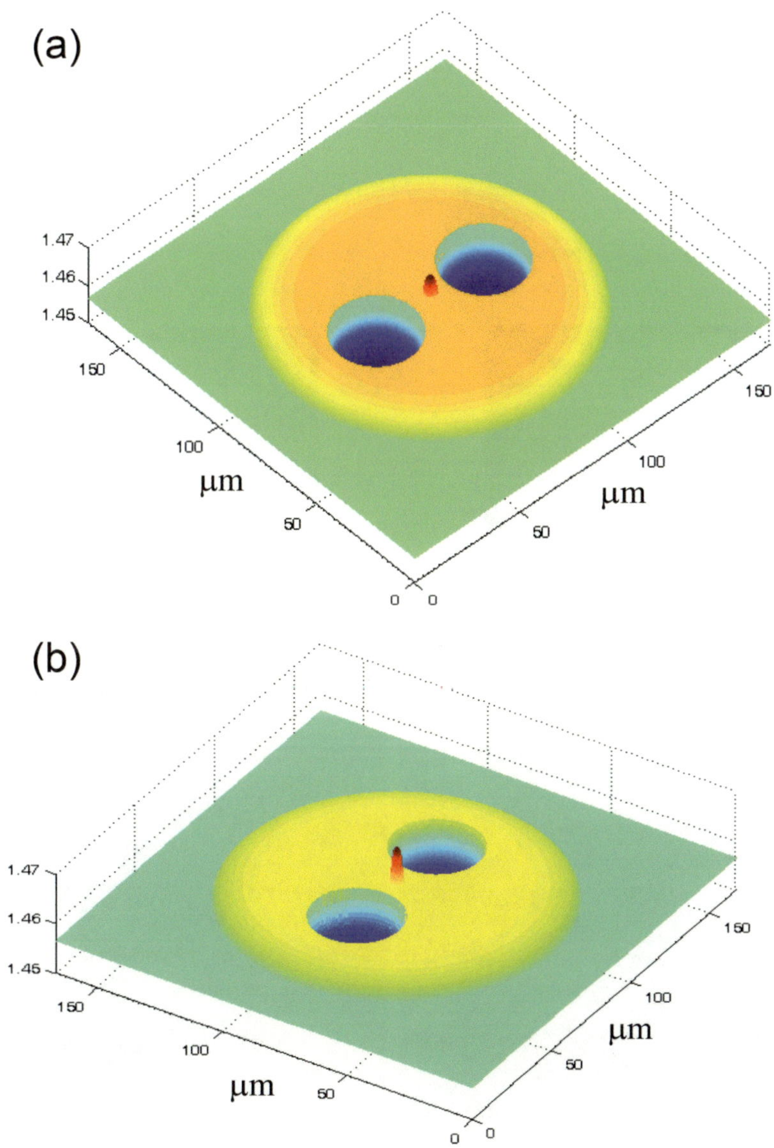

Figure 18.
The 3D RIPs of PANDA PM optical fiber in the directions of (a) fast axis and (b) slow axes. Ref. [23] with permission.

4.1 Panda optical fiber

PANDA PM optical fiber is preferable in telecommunications [57, 58]. It is modified by insertion of stress rods to provide PM properties according to the procedure described in **Figure 16**. In this process, two holes are ultrasonically

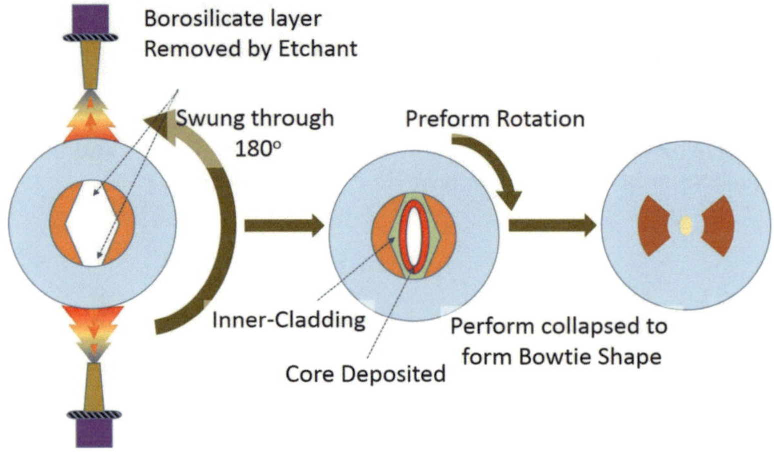

Figure 19.
Manufacturing steps of a bow tie PM optical fiber.

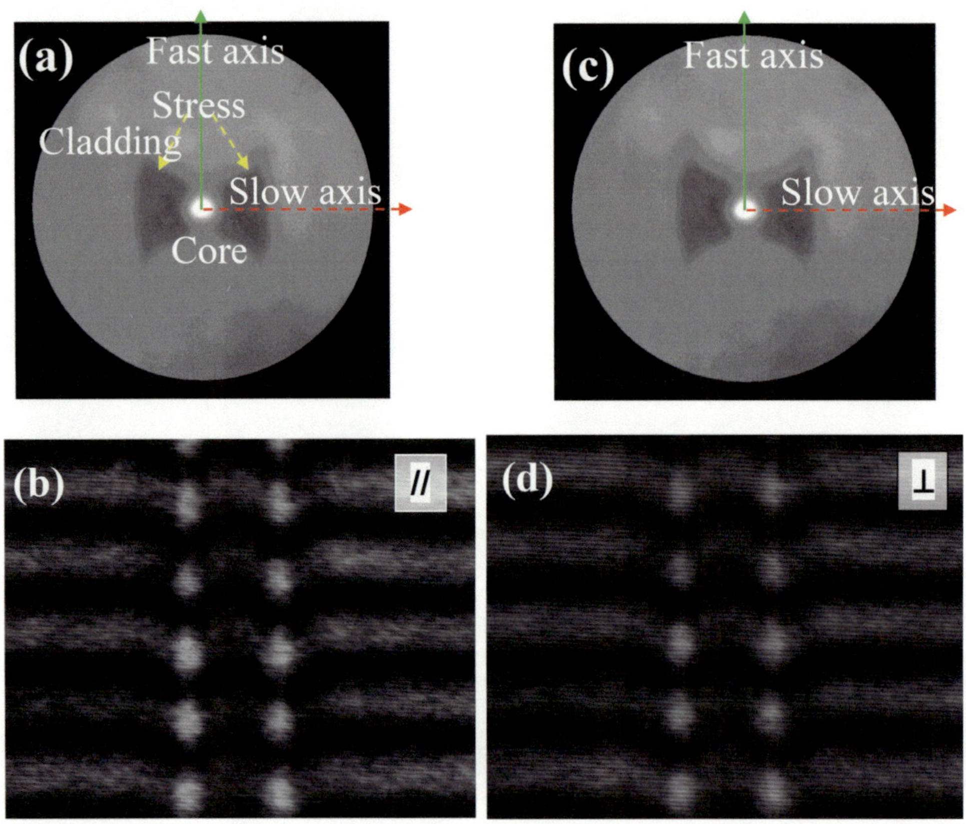

Figure 20.
(a and c) Cross-sections of the bow tie optical fiber. (b and d) Experimentally obtained phase shifted interferograms when the incident light vibrates parallel and perpendicular to fiber's axis, respectively. Ref. [59] with permission.

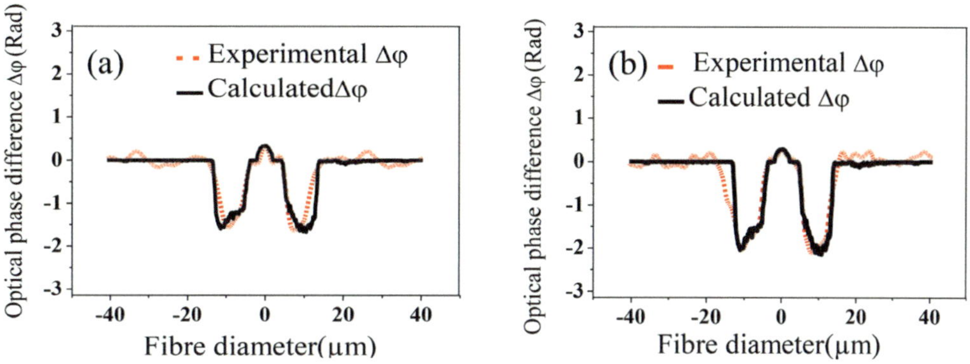

Figure 21.
The calculated and the experimental phase differences of the bow tie optical fiber when the incident light vibrates (a) parallel and (b) perpendicular to the fiber's axis. Ref. [59] with permission.

drilled along a single-mode optical fiber; then, the stress rods are inserted in these two holes and the fiber is finally drawn [57]. In 2014, Wahba used the off-axis DHPSI to reconstruct the 3D RIP of a PM PANDA optical fiber [23]. The multilayer model was used to calculate the RIP of this fiber in the directions of fast and slow axes. By rotating the PANDA fiber, different interferograms were recorded and analyzed in order to reconstrut the 3D RIP of this fiber, see **Figure 17**. The reconstructed 3D RIPs of PANDA fiber are shown in **Figure 18** when the incident light was vibrating in the direction of (a) fast axis and (b) slow axis.

4.2 Bow tie optical fiber

A bow tie optical fiber is fabricated on a lathe using the inside vapor-phase oxidation (IVPO) via the process called gas-phase etching to create the required stress [57]. This process is summarized in **Figure 19** where a ring of boron-doped silica is purely deposited of boron tribromide in combination with silicon tetrachloride. The rotation of the lathe stopped when a sufficiently thick layer was formed to allow two diametrically opposed sections to be etched away. The final shape of the bow tie and stress levels are controlled by varying the arc through which the etching burner is rotated. Recently, Ramadan et al. estimated the optical phase variations of optical rays traversing a PM optical fiber from its cross-section images [59]. They proposed an algorithm to recognize the different areas of the fiber's cross-section, which was immersed in a matching liquid and investigated by Mach-Zehnder interferometer.

These areas were scanned to calculate the optical paths for certain values of refractive indices and the optical phases across the PM optical fiber were recovered. The experimental interferograms of the bow tie PM optical fiber, shown in **Figure 20**, were analyzed to extract their optical phase distributions and compare them with the optimized estimated optical phase maps, see **Figure 21**. This was a direct and accurate method to get information about refractive index, birefringence, and the beat length of a PM optical fiber.

5. Homogeneous thick optical fibers

Optical fibers having diameters in the order of 100 μm, or less, are convenient to be investigated using interferometric methods when the samples are put in immersion liquids of refractive indices close to the refractive indices of the fibers as described in the previous sections [12, 13]. Optical fibers of diameters bigger than 150 μm cannot be investigated by normal interferometry where the planes of fringes in both liquid and fiber cannot be focused simultaneously. In 2000, Ramadan presented a novel interferometric method to recover such a problem for homogeneous thick optical fibers, commonly used in short-distance data transmission, without using immersion liquids [16]. This type of interference was called lens-fiber interferometry (LFI) since the interference fringes were produced by a combination of an aberrated cylindrical lens and a thick optical fiber. The aberrated cylindrical lens was used to focus a parallel beam on this fiber, which was located in the focal plane of the cylindrical lens [60], see **Figure 22**.

Two-beam interference produced by the superposition of two optical rays emerging from the fiber was recorded and explained. Due to the aberration of the cylindrical lens, one of these two rays crossed the thick fiber before its center while the other ray crossed after the fiber's center. Therefore, for each point in the image plane, two rays having two different initial incidence angles on the thick fiber are superposed, see **Figure 23**. The optical path length of each ray can be obtained by tracing this ray geometrically, as given by Eq. (30), which can be transformed into phase differences for the interfered rays using Eq. (31). The difference in the optical path lengths of each pair of interfered rays can be transformed into an intensity distribution describing the interference fringes using Eq. (32). On the other hand, the scattered rays from the outer surface of the fiber do not contribute in the interference because of the limited range of the incident rays on the fiber. This is in contrast with previous works done by Watkins [14, 15, 61]. By comparing the experimentally obtained interferograms with those reconstructed theoretically as

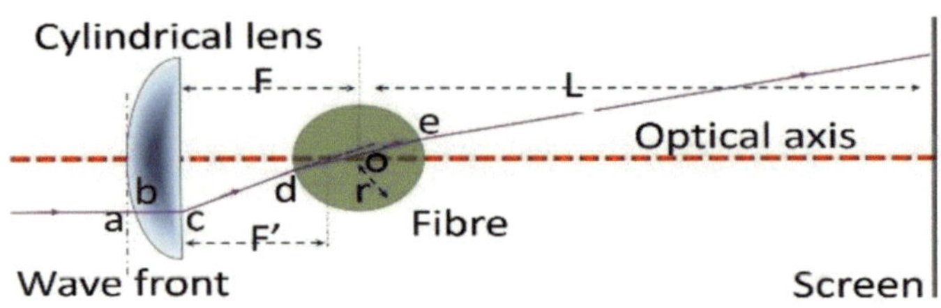

Figure 22.
The ray tracing diagram of an optical ray crossing a homogeneous thick optical fiber.

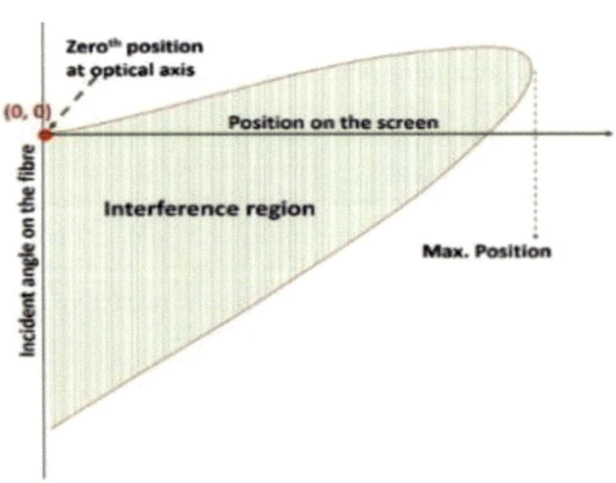

Figure 23.
The relation between the position of each two interfered rays on the screen and their incidence angles on the thick fiber.

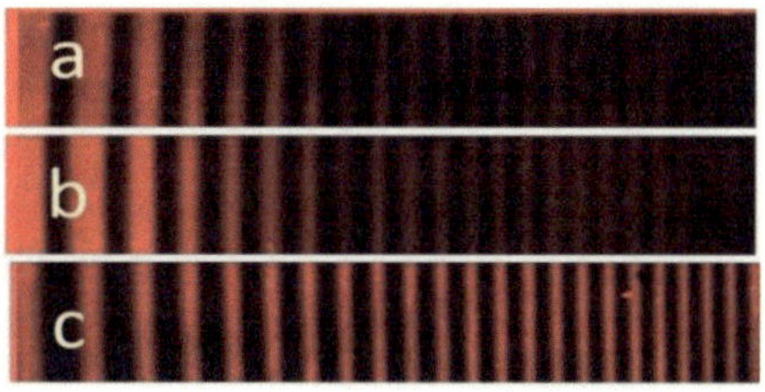

Figure 24.
(a) A selected and extended part of the obtained interferogram of a thick optical fiber, (b) the enhanced fringes of (a) and (c) the simulated fringes.

shown in **Figure 24**, Ramadan was able to determine the refractive index of the investigated thick optical fiber. The advantage is that the used system requires no matching liquid where the experiment is performed when the thick fiber is just held in air. This enables monitoring the probable variation in radius or refractive index of the fiber particularly during the manufacturing process or under external effects.

$$\Delta(\mathrm{z}) = ab + bc \cdot n_L + cd + de \cdot n_f + ep \tag{30}$$

$$\delta = \frac{2\pi}{\lambda} \cdot (\Delta(z_1) - \Delta(z_2)) \tag{31}$$

$$I = 4A^2 \cos^2\left(\frac{\delta}{2}\right) \tag{32}$$

where $\Delta(\mathrm{z}_1)$ and $\Delta(\mathrm{z}_2)$ are the optical path lengths of the two interfered rays. In 2004, Hamza et al. developed LFI technique in order to determine the refractive index of the core of a skin-core thick optical fiber [60]. They derived a mathematical expression for the optical paths through the fiber in order to reconstruct the interfernce pattern due to the used fiber when it is used as a thick fiber in the LFI system. By comparing the experimentally obtained patterns with the theoretically reconstructed ones, they were able to estimate the core's refractive index with an accuracy of 8×10^{-4}. Due to its simplicity and applicability, LFI was used, afterward, to measure the refractive index of a liquid [62] and to monitor the thickness variations of a transparent sheet inserted between the cylindrical lens and the thick fiber [63].

6. Conclusions

This chapter is an attempt to highlight the interferometric techniques used for characterization of optical fibers. Application of two- and multiple-beam interference on different types of fibers is illustrated. Section 2 dealt with conventional optical fibers where we illustrated the theoretical models used to reconstruct the refractive index profiles of these fibers. In these models, the refraction of the light ray traversing the fiber has been considered. Digital holography was explained as an important candidate used for accurate retrieving of phase maps and consequently refractive index profiles of the fibers. In Section 3, we mentioned the problem of fiber bending. Recovering the refractive index profile and mode propagation of a bent fiber considering the refraction of the light rays traversing the fiber is a quite difficult task since bending-induced stresses are responsible for refractive index variations. Also, these stresses are released at the outer surface of the bent fiber. Therefore, we illustrated a successful model that was recently presented to recover the index profile in this case with experimental illustrative data. Another important type of optical fibers is the polarization maintaining optical fibers, which prevent cross-coupling by conserving the state of beam polarization during propagation. In Section 4, we presented interferometric techniques applied on two different polarization maintaining optical fibers, panda and bow tie, to reconstruct their refractive index profiles. Most interference techniques require immersing the fiber in a suitable liquid in order to minimize the phase difference between the fiber and its surrounding medium. In Section 5, an interference technique is presented and applied on a thick optical fiber to recover its refractive index without using an immersion liquid (i.e., in air), which makes the technique suitable for in-situ studying of thick fibers.

Acknowledgements

The authors would like to acknowledge Prof. A. Hamza, the leader of optics research groups in Mansoura and Damietta Universities, and Prof. T. Sokkar for their continuous support and useful discussions. Also, many thanks to the Optics Research Group members in Damietta University for their useful suggestions and comments.

Conflict of interest

The authors declare no conflict of interest.

Author details

Hamdy Wahba[1,2]*, Wael Ramadan[1] and Ahmed El-Tawargy[1]

1 Physics Department, Faculty of Science, Damietta University, New Damietta, Egypt

2 Physics Department, Faculty of Science, Taif University, Taif, Saudi Arabia

*Address all correspondence to: hhwahba76@gmail.com

References

[1] El-Diasty F. Characterization of optical fibers by two- and multiple-beam interferometry. Optics and Lasers in Engineering. 2008;**46**:291-305. DOI: 10.1016/j.optlaseng.2007.10.004

[2] Midwinter JE. Optical Fibers for Transmission. New York: Wiley; 1979

[3] Barakat N, Hamza AA. Interferometry of Fibrous Materials. Bristol, New York: A. Hilger; 1990

[4] DeCusatis C. Optical data communication: Fundamentals and future directions. Optical Engineering. 1998;**37**:3082. DOI: 10.1117/1.601993

[5] Udd E, Spillman WB, editors. Fiber Optic Sensors: An Introduction for Engineers and Scientists. Hoboken, NJ, USA: John Wiley & Sons, Inc.; 2011. DOI: 10.1002/9781118014103

[6] Hall DR, Jackson PE, editors. The Physics and Technology of Laser Resonators. Bristol, New York: A. Hilger; 1989

[7] Erdogan T. Fiber grating spectra. Journal of Lightwave Technology. 1997; **15**:1277-1294. DOI: 10.1109/50.618322

[8] Meltz G, Morey WW, Glenn WH. Formation of Bragg gratings in optical fibers by a transverse holographic method. Optics Letters. 1989;**14**:823. DOI: 10.1364/OL.14.000823

[9] Hill KO, Malo B, Vineberg KA, Bilodeau F, Johnson DC, Skinner I. Efficient mode conversion in telecommunication fibre using externally written gratings. Electronics Letters. 1990;**26**:1270. DOI: 10.1049/el:19900818

[10] De Paula RP, Moore EL. In: Moore EL, Ramer OG, editors. Review of All-Fiber Phase and Polarization Modulators. Arlington, United States: SPIE. Digital library, Technical Symposium East; 1984. p. 3. DOI: 10.1117/12.942649

[11] El-Sherif MA. On-fiber sensor and modulator. IEEE Transactions on Instrumentation and Measurement. 1989;**38**:595-598. DOI: 10.1109/19.192356

[12] Hamza AA, Ghander AM, Sokkar TZN, Mabrouk MA, Ramadan WA. On the determination of the refractive index of a fibre: I. Skin-core fibre. Pure and Applied Optics. 1994;**3**:943-961. DOI: 10.1088/0963-9659/3/6/003

[13] Hamza AA, Sokkar TZN, Ghander AM, Mabrouk MA, Ramadan WA. On the determination of the refractive index of a fibre. II. Graded index fibre. Pure and Applied Optics. 1995;**4**:161-177. DOI: 10.1088/0963-9659/4/3/004

[14] Watkins LS. Scattering from side-illuminated clad glass fibers for determination of fiber parameters. Journal of the Optical Society of America. 1974;**64**:767. DOI: 10.1364/JOSA.64.000767

[15] Watkins LS. Laser beam refraction traversely through a graded-index preform to determine refractive index ratio and gradient profile. Applied Optics. 1979;**18**:2214. DOI: 10.1364/AO.18.002214

[16] Ramadan WA. On-line lens-fibre interference method for testing a thick fibre. Journal of Optics A: Pure and Applied Optics. 2000;**2**: 234-238. DOI: 10.1088/1464-4258/2/3/311

[17] Wahba HH, Kreis T. In: Lehmann PH, editor. Characterization

of Optical Fibers by Digital Holographic Interferometry. Munich, Germany: SPIE. Digital library, Europe Optical Metrology; 2009. p. 73890K. DOI: 10.1117/12.826854

[18] Wahba HH, Kreis T. Digital holographic interferometric characterization of bent optical fibers. Journal of Optics A: Pure and Applied Optics. 2009;**11**:105407. DOI: 10.1088/1464-4258/11/10/105407

[19] Wahba H. Digital Holography and Interferometric Metrology of Optical Fibres Digital Holographic Phase Shifting and Interferometric Characterization of Optical Fibers. VDM Verlag Dr. Müller: Saarbrücken; 2011

[20] El-Din MAS, Wahba HH. Investigation of refractive index profile and mode field distribution of optical fibers using digital holographic phase shifting interferometric method. Optics Communications. 2011;**284**:3846-3854. DOI: 10.1016/j.optcom.2011.04.037

[21] Wahba H, Shams M. Digital holographic interferometric characterization of optical waveguides. In: Naydenova I, editor. Advanced Holography - Metrology and Imaging. InTech; 2011. DOI: 10.5772/22311

[22] Sokkar TZN, Ramadan WA, Shams El-Din MA, Wahba HH, Aboleneen SS. Bent induced refractive index profile variation and mode field distribution of step-index multimode optical fiber. Optics and Lasers in Engineering. 2014;**53**:133-141. DOI: 10.1016/j.optlaseng.2013.09.002

[23] Wahba HH. Reconstruction of 3D refractive index profiles of PM PANDA optical fiber using digital holographic method. Optical Fiber Technology. 2014;**20**:520-526. DOI: 10.1016/j.yofte.2014.06.002

[24] Sokkar TZN, El-Farahaty KA, Ramadan WA, Wahba HH, Raslan MI, Hamza AA. Nonray-tracing determination of the 3D refractive index profile of polymeric fibres using single-frame computed tomography and digital holographic interferometric technique: Single frame computed tomography. Journal of Microscopy. 2015;**257**:208-216. DOI: 10.1111/jmi.12203

[25] Ramadan WA, El-Tawargy AS. A double image Mach–Zehnder interferometer. Applied Physics B. 2015;**118**:203-208. DOI: 10.1007/s00340-014-5970-3

[26] Wahba HH, Kreis T. Characterization of graded index optical fibers by digital holographic interferometry. Applied Optics. 2009;**48**:1573. DOI: 10.1364/AO.48.001573

[27] Yassien KM, Agour M, von Kopylow C, El-Dessouky HM. On the digital holographic interferometry of fibrous material, I: Optical properties of polymer and optical fibers. Optics and Lasers in Engineering. 2010;**48**:555-560. DOI: 10.1016/j.optlaseng.2009.12.003

[28] Wahba HH, Shams El-Din MA. In: Lehmann PH, Osten W, Gastinger K, editors. Reconstruction of 3D refractive index distribution across the graded index optical fibre using digital holographic interferometry. Munich, Germany; 2011. p. 80822B. DOI: 10.1117/12.882164

[29] Wahba HH, Sjödahl M, Gren P, Olsson E. High resolution digital holographic microscopy for the study of aggregated natural cellulose nanowhisker fibers. Optics and Lasers in Engineering. 2015;**73**:69-74. DOI: 10.1016/j.optlaseng.2015.04.005

[30] Sokkar TZN, Shams El-Din MA, El-Tawargy AS. Effect of recycling on the optical, mechanical and structural properties of polypropylene fibers. Optics and Lasers in Engineering. 2013;**51**:994-1003. DOI: 10.1016/j.optlaseng.2013.02.018

[31] El-Morsy MA, Yatagai T, Hamza A, Mabrouk MA, Sokkar TZN. Multiple-beam Fizeau fringe-pattern analysis using Fourier transform method for accurate measurement of fiber refractive index profile of polymer fiber. Journal of Applied Polymer Science. 2002;**85**: 475-484. DOI: 10.1002/app.10387

[32] Shams El-Din MA, El-Tawargy AS. Interferometric study of the effect of laser intensity and polarization on the cold-drawing of virgin polypropylene fibres. Indian Journal of Physics. 2017; **91**:1425-1435. DOI: 10.1007/s12648-017-1044-9

[33] Ramadan WA, Wahba HH, Shams El-Din MA. Two-dimensional refractive index and stresses profiles of a homogenous bent optical fiber. Applied Optics. 2014;**53**:7462. DOI: 10.1364/AO.53.007462

[34] Sokkar TZN, Shams El-Din MA, El-Tawargy AS. On Young's modulus profile across anisotropic nonhomogeneous polymeric fibre using automatic transverse interferometric method. Optics and Lasers in Engineering. 2012;**50**:1223-1229. DOI: 10.1016/j.optlaseng.2012.03.017

[35] Ramadan WA, Wahba HH, Shams El-Din MA. Two-dimensional refractive index and birefringence profiles of a graded index bent optical fibre. Optical Fiber Technology. 2017;**36**:115-124. DOI: 10.1016/j.yofte.2017.03.005

[36] Hamza AA, Mabrouk MA. A multiple-beam interferometric method for refractive index determination of graded optical fibres. Journal of Modern Optics. 1991;**38**:97-107. DOI: 10.1080/09500349114550131

[37] Hamza AA, Sokkar TZN, El-Morsy MA, Nawareg MAE. Automatic determination of refractive index profile of fibers having regular and/or irregular transverse sections considering the refraction of light rays by the fiber. Optics Communications. 2009;**282**: 27-35. DOI: 10.1016/j.optcom.2008.09.073

[38] Hamza AA, Sokkar TZN, El-Morsy MA, Nawareg MAE. Automatic determination of refractive index profile, sectional area, and shape of fibers having regular and/or irregular transverse sections. Optics & Laser Technology. 2008;**40**:1082-1090. DOI: 10.1016/j.optlastec.2008.01.022

[39] Ramadan WA. Intensity distribution of Fizeau fringes in transmission with the real path of the interfered multiple-beams. Optics and Lasers in Engineering. 2014;**58**:27-32. DOI: 10.1016/j.optlaseng.2014.01.017

[40] Born M, Wolf E. Principles of Optics: Electromagnetic Theory of Propagation, Interference and Diffraction of Light. 7th Expanded Ed. Cambridge. New York: Cambridge University Press; 1999

[41] Gabor D. A new microscopic principle. Nature. 1948;**161**:777-778. DOI: 10.1038/161777a0

[42] Microscopy by reconstructed wave-fronts. Proceedings of the Royal Society of London A. 1949;**197**:454-487. DOI: 10.1098/rspa.1949.0075

[43] Gabor D. Microscopy by reconstructed wave fronts: II. Proceedings of the Physical Society. Section B. 1951;**64**:449-469. DOI: 10.1088/0370-1301/64/6/301

[44] Powell RL, Stetson KA. Interferometric vibration analysis by Wavefront reconstruction. Journal of the Optical Society of America. 1965;**55**: 1593. DOI: 10.1364/JOSA.55.001593

[45] Kreis T. Holographic Interferometry: Principles and Methods. Berlin: Akademie Verlag; 1996

[46] Schnars U. Direct phase determination in hologram

interferometry with use of digitally recorded holograms. Journal of the Optical Society of America. A. 1994;**11**:2011. DOI: 10.1364/JOSAA.11.002011

[47] Schnars U. Digital recording and numerical reconstruction of holograms: Reduction of the spatial frequency spectrum. Optical Engineering. 1996;**35**:977. DOI: 10.1117/1.600706

[48] Schnars U, Jueptner W. Digital Holography: Digital Hologram Recording, Numerical Reconstruction, and Related Techniques. Berlin: Springer; 2005

[49] Hariharan P, Oreb BF, Brown N. Real-time holographic interferometry: A microcomputer system for the measurement of vector displacements. Applied Optics. 1983;**22**:876. DOI: 10.1364/AO.22.000876

[50] Hamza AA, Mabrouk MA, Ramadan WA, Shams-Eldin MA. Determination of GR-IN optical fibre parameters from transverse interferograms considering the refraction of the incident ray by the fibre. Optics Communications. 2001;**200**:131-138. DOI: 10.1016/S0030-4018(01)01561-9

[51] Mabrouk MA. Two-beam interference detection of the changes in fibre structural parameters during low drawing process. Polymer Testing. 2002;**21**:897-904. DOI: 10.1016/S0142-9418(02)00026-0

[52] Fan Y, Wu G, Wei W, Yuan Y, Lin F, Wu X. Fiber-optic bend sensor using LP_21 mode operation. Optics Express. 2012;**20**:26127. DOI: 10.1364/OE.20.026127

[53] Méndez A, Morse TF. Specialty Optical Fibers Handbook. Amsterdam [u.a.]: Elsevier/Academic Press; 2007

[54] Yin S, Ruffin PB, Yu FTS. Fiber Optic Sensors. S.l.: CRC Press; 2019

[55] Vogler DE, Lorencak A, Rey JM, Sigrist MW. Bending loss measurement using a fiber cavity ringdown scheme. Optics and Lasers in Engineering. 2005;**43**:527-535. DOI: 10.1016/j.optlaseng.2004.03.012

[56] Makouei S, Oskouei MS, Rostami A. Study of bending loss and mode field diameter in depressed inner core triple-clad single-mode optical fibers. Optics Communications. 2007;**280**:58-67. DOI: 10.1016/j.optcom.2007.07.052

[57] Emslie C. Polarization maintaining fibers. In: Specialty Optical Fibers Handbook. Elsevier; 2007. pp. 243-277. DOI: 10.1016/B978-012369406-5/50010-2

[58] Hosaka T, Okamoto K, Miya T, Sasaki Y, Edahiro T. Low-loss single polarisation fibres with asymmetrical strain birefringence. Electronics Letters. 1981;**17**:530. DOI: 10.1049/el:19810371

[59] Ramadan WA, Wahba HH, Shams El-Din MA, Abd El-Sadek IG. Refractive index retrieving of polarization maintaining optical fibers. Optical Fiber Technology. 2018;**40**:69-75. DOI: 10.1016/j.yofte.2017.11.007

[60] Hamza AA, Mabrouk MA, Ramadan WA, Wahba HH. Core-index determination of a thick fibre using lens-fibre interference (LFI) technique. Optics and Lasers in Engineering. 2004;**42**:121-130. DOI: 10.1016/j.optlaseng.2003.09.002

[61] Watkins LS. Effect of noncircular cross section on the forwardscattering pattern of side-illuminated unclad fibers. Applied Optics. 1979;**18**:4089. DOI: 10.1364/AO.18.004089

[62] Ramadan WA, Wahba HH. Lens–fibre interference in measuring liquids' refractive indices. Measurement Science and Technology. 2006;**17**:215-220. DOI: 10.1088/0957-0233/17/1/033

[63] Ramadan WA, Shams El-Din MA, Wahba HH, El-Tawargy AS, Hamza AA. Lens–fibre interference proposed to monitor a transparent sheet's thickness variations. Applied Physics B: Lasers and Optics. 2014;**117**:1073-1080. DOI: 10.1007/s00340-014-5928-5

Section 4

Optical Fiber Sensors

Chapter 6

Temperature Sensing Characteristics of Tapered Doped Fiber Amplifiers

Rafael Sanchez-Lara, Lelio de la Cruz-May,
José Luis Vazquez-Avila,
Daniel Enrique Ceballos-Herrera and
José Alfredo Alvarez-Chavez

Abstract

We numerically analyze the temperature response of tapered doped fiber amplifiers and discuss their feasibility to be used as a sensing element in temperature fiber sensors. In particular, we consider Ytterbium (Yb) and Thulium (Tm) rare earths in the tapered doped fiber designs. We have modified the coupled propagation equations for the pump and signal radiations in order to include different taper structures and introduce the temperature dependence of the absorption and emission cross-sections of Yb and Tm ions. It was found that the temperature sensitivity of the amplified signal in Tm-doped fiber amplifiers is one order of magnitude higher than this obtained with Yb-doped fibers. Additionally, in all doped fibers, the temperature sensitivity of the signal radiation is higher for low pump powers in a co-propagating pump scheme, and it highly depends on the longitudinal shape of the taper used. Finally, for both Yb- and Tm-doped fibers, the temperature sensitivity can be increased if we use doped fiber lengths shorter than 1 m and pump powers lower than 300 mW. This study provides valuable information for the development of tapered fiber amplifiers doped with other rare earths and novel designs for doped fiber temperature sensors.

Keywords: ytterbium, thulium, taper, doped fiber, sensors

1. Introduction

Non-invasive sensors for temperature and strain measurements in explosive enviroments and with immunity to electromagnetic interference are constantly required. In this context, optical fibers have become a key piece to develop temperature sensors in such hazardous applications. As an example of these applications, we can mention petroleum pipeline monitoring and hydraulic fracturing evaluations [1, 2]. Consequently, many efforts have been made to improve the performance of fiber optic temperature sensors based on different principles and designs.

IntechOpen

The development of temperature and strain fiber sensors for environmental measurements where immunity to electromagnetic interference and high personal safety are required has been studied with great interest until today. This fact has allowed the development of a large number of fiber optic temperature sensors based in different principles and desgins; among them, we can distinguish two principal groups which are based in doped [3–10] and un-doped fibers [11–17]. In the first group, we have sensors that use the temperature dependence of the fluorescence lifetime and involve techniques as the fluorescence intensity ratio. In these sensors, the key parameter is the temperature dependence of the absorption and emission cross-sections of the pump and signal in the doped fiber amplifier [18–22]. On the other hand, the second group of temperature sensors are based on un-doped fibers, and these use principally interference techniques. As examples, we can mention sensors based on Fabry-Perot interferometers, fiber Bragg gratings, long-period gratings, optical fiber couplers, and tapered fibers [11–17]. The key parameter in these sensors is the temperature dependence of the dielectric material that modifies the modal behavior of the signal that propagates in the un-doped fiber. It consequently changes the interferometer condition and generates a power variation at the end of the fiber device. Currently, several works have been performed to improve the sensitivity of these temperature sensors employing a combination of these sensing techniques in doped and un-doped fibers, respectively. Additionally, these combinations have allowed discerning between simultaneous strain and temperature measurements [3–6]. In this context, the gradual progress to develop improved temperature fiber sensors requires the necessity to explore continuously novel configurations based on the sensing techniques described above. At this point, one interesting proposal is to consider the use of a tapered fiber, amply used as a temperature fiber sensor [11–17], inscribed simultaneously in a doped fiber amplifier, which poses an additional temperature response caused by its cross-sections [3–10, 18–21]. At this respect, only the efficiency of pump absorption in tapered doped fiber lasers has been studied [23, 24], but a detailed analysis of its temperature response has not be performed. Therefore, in this chapter, we present an analysis of the temperature sensing characteristics of a tapered Yb- and Tm-doped fiber amplifier, and we explore its feasibility to be used as sensing element in temperature fiber sensors. In the present analysis, we study the thermal response of the tapered doped fiber with different tapered fiber structures and pump schemes to find an optimized design for temperature sensing.

2. Numerical simulation

We start our analysis considering first a tapered fiber section formed in a single-mode step index fiber surrounded by air. This scheme has been extensively studied in un-doped fibers showing that no changes of the transmitted power by temperature are obtained when only air is considered as surrounding media [11–14]. These modifications on the transmitted power are reached when a thermochromic material surrounds the tapered fiber section. However, this situation is not considered in our study because our principal interest is to analyze the transmitted power variations caused by the temperature response of the amplified spontaneous emission in the tapered doped core.

Once air is considered as the surrounding media, we proceed to analyze the mode field expression of the fundamental core mode within the tapered fiber. The mode field expression is an important parameter in doped fiber amplifiers because it determines the core mode fraction of the pump and signal, which affects the power conversion in the doped core [23–28]. In this study, Gaussian shapes for the pump and signal fundamental core modes were considered. Thus, we can write the transverse intensity pattern using a Gaussian envelope approximation given by [29]:

$$f_{p,s}(r) = \frac{1}{\pi\Omega^2} e^{-r/\Omega^2} \tag{1}$$

where subscripts p and s refer to pump and signal radiations and Ω is determined by the characteristic of the fiber: the refractive indexes and radius of the core and cladding, respectively. The multiplying factor in Eq. (1) is chosen to normalize $f(r)$ as follows:

$$2\pi \int f_{p,s}(r)dr = 1 \tag{2}$$

For a step index fiber, Ω is approximately given by [29]:

$$\Omega = aJ_0(U) \frac{V}{U} \frac{K_1(W)}{K_2(W)} \tag{3}$$

where a is the core radius and $U = a\sqrt{k^2 n_1^2 - \beta^2}$, $W = a\sqrt{\beta^2 - k^2 n_1^2}$, $V = ka\sqrt{n_1^2 - n_2^2}$, $k = \frac{2\pi}{\lambda}$, n_1 and n_2 are the refractive indices of the active core and cladding, respectively, and β is the propagation constant of the pump or signal mode. For a given V, the value of W can be obtained using the empirical relationship

$$W = 1.1428V - 0.996, \tag{4}$$

It is valid only for step index fibers. In this way, according to the pump and signal wavelengths, one can obtain the respective values of U, W, and V using the numerical aperture (NA) of the fiber where

$$NA = \sqrt{n_1^2 - n_2^2}, \tag{5}$$

and subsequently the parameter Ω. This procedure is accurate for $1.5 < V < 2.5$.

Now, we analyze the thermal effects on the tapered doped fiber considering the changes of the absorption and emission cross-section with temperature. Previous works in un-tapered fibers have reported changes in population of the energy levels in Yb-doped glasses, and the broadening of the homogeneous linewidth as temperature is increased [18–21]. It results in modifications of the absorption and emission cross-sections for the signal and pump radiations. These effects may vary for each individual doped fiber due to the different glass composition, concentration of dopants and co-dopants, and the degree of structural disorder on the glass network used in different active fibers. In particular, we study the impact of variations on the cross-sections due to temperature in tapered doped fiber amplifiers. Without loss of generality, we use the absorption and emission cross-section changes with temperature reported in [18], which is a representative of several Yb-doped fibers. These changes are expressed in the following equations:

$$\sigma(T) = \sigma(20^\circ C) + \frac{d\sigma}{dT} \tag{6}$$

$$\frac{d\sigma_{abs}^{1064nm}}{dT} = 7.78 * 10^{-29} \frac{m^2}{^\circ K} \tag{7}$$

$$\frac{d\sigma_{em}^{1064nm}}{dT} = -2.44 * 10^{-28} \frac{m^2}{^\circ K} \tag{8}$$

$$\frac{d\sigma_{abs}^{976nm}}{dT} = \frac{d\sigma_{em}^{976nm}}{dT} = -1.63 * 10^{-27} \frac{m^2}{^\circ K} \tag{9}$$

where σ_{abs}^{1064nm} and σ_{em}^{1064nm} are the absorption and emission cross-sections for the signal wavelength and σ_{abs}^{976nm} and σ_{em}^{976nm} are the absorption and emission cross-sections for the pump wavelength, respectively.

In order to model a temperature-dependent tapered Yb-doped fiber amplifier, we numerically analyze the following coupled equations:

$$\frac{dI_p(r,z)}{dz} = \left[\sigma_{em}^{p}(T)n_2(T) - \sigma_{abs}^{p}(T)n_1(T)\right]N_{tot}I_p(r,z) \tag{10}$$

$$\frac{dI_s(r,z)}{dz} = \left[\sigma_{em}^{s}(T)n_2(T) - \sigma_{abs}^{s}(T)n_1(T)\right]N_{tot}I_s(r,z) \tag{11}$$

where $I_p(r,z)$ and $I_s(r,z)$ are the pump and signal intensities; N_{tot} is the total ytterbium population; σ_{abs}^{p}, σ_{em}^{p}, σ_{abs}^{s}, σ_{em}^{s}, are the temperature dependent absorption and emission cross-sections of the pump and signal at 976 nm and 1064 nm, respectively; and $n_1(T)$, $n_2(T)$ are the temperature-dependent upper- and lower-state populations of Yb, which are given at a steady state by the following equations:

$$n_2 = \frac{R_{abs} + W_{abs}}{R_{abs} + R_{em} + W_{abs} + W_{em} + A_{esp}} \tag{12}$$

$$n_1 = 1 - n_2 \tag{13}$$

where $R_{abs} = \sigma_{abs}^{p} I_p h v_p$, $R_{em} = \sigma_{em}^{p} I_p h v_p$, $W_{abs} = \sigma_{abs}^{s} I_s h v_s$, and $W_{em} = \sigma_{em}^{s} I_s h v_s$. In these equations, the ASE generation is not considered, and only effects of the taper and temperature modifications are investigated.

In order to consider the taper effects and the overlap of the pump and signal fundamental mode with the active core, we can use [29]:

$$I_{p,s} = P_{p,s}(z) f_{p,s}(r) \tag{14}$$

where subscripts p and s refer to pump and signal radiations, $P_{p,s}(z)$ are the z-dependent powers at the pump and signal wavelengths, and $f(r)$ is given by Eq. (1). It is clear at this point that the effect of taper and temperature in parameter Ω defined in Eq. (3) directly modifies the evolution of pump and signal intensities described in Eqs. (10) and (11).

If we consider the pump power at any value of z, we have [30]

$$P_{p,s}(z) = \int_0^\infty \int_0^{2\pi} I_{p,s}(r,z) r dr d\phi = 2\pi \int_0^\infty I_{p,s}(r,z) r dr \tag{15}$$

Then,

$$\frac{dP_{p,s}(z)}{dz} = 2\pi \int_0^\infty \frac{dI_{p,s}(r,z) r dr}{dz} \tag{16}$$

Using Eqs. (14) and (16), we can rewrite Eqs. (10) and (11) as follows:

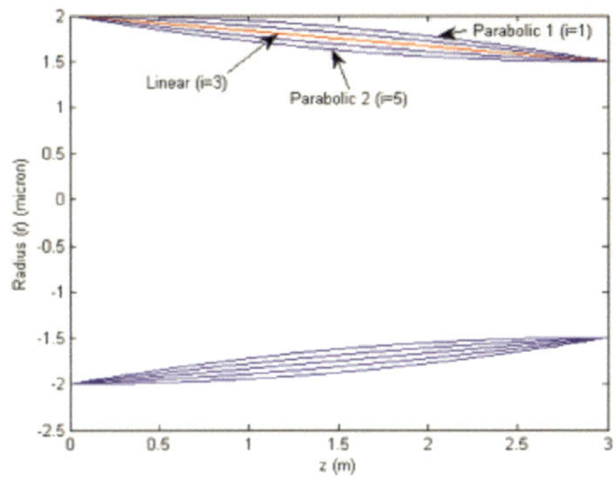

(a)

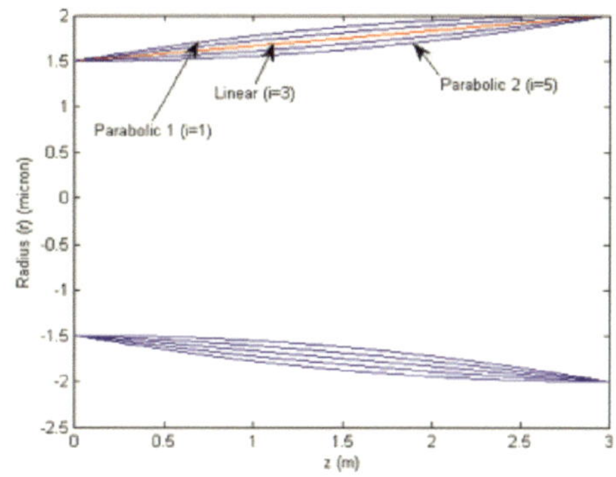

(b)

Figure 1.
Modeling scheme using tapered Yb-doped amplifiers with different tapered core shapes and with co-propagating single pump at different initial taper ends: (a) pump at the end with wider radius and (b) pump at the end with lower radius.

$$\frac{dP_p(z)}{dz} = 2\pi \int_0^{a(z)} \left[\sigma_{em}^p(T) n_2(T) - \sigma_{abs}^p(T) n_1(T)\right] N_{tot} f_p(r) r dr \tag{17}$$

$$\frac{dP_s(z)}{dz} = 2\pi \int_0^{a(z)} \left[\sigma_{em}^s(T) n_2(T) - \sigma_{abs}^s(T) n_1(T)\right] N_{tot} f_s(r) r dr \tag{18}$$

On these equations, we assume that the fiber is doped with uniform Yb concentration up to the core radius "a" which depends on z. Besides, it is worth to note that $n_1(T)$ and $n_2(T)$ depend on $f_{p,s}(r)$ due to their relation with I_p and I_s intensities.

Once the temperature-dependent coupled equations are defined, we proceed to model a tapered Yb-doped fiber amplifier with different longitudinal tapered core shapes and different pump schemes [31] as is shown in **Figure 1**.

$$r(z) = \frac{1}{C}\left(-D - Bz - z^2\right) \tag{19}$$

$$r(z) = \frac{1}{C}\left(-D - B(L - z) - (L - z)^2\right) \tag{20}$$

This model can be applied to a different doped material as the thulium; in this case Eqs. (6) to (9) must be replaced in agreement with absorption and emission cross-section changes with temperature reported in [32–35]. These changes are expressed in the following equations:

$$\sigma_{abs}^{1600nm}(T) = 15.56x10^{-25} - 38x10^{-28}T \tag{21}$$

$$\sigma_{em}^{1600nm}(T) = 1.8x10^{-26} + 1.99x10^{-28}T \tag{22}$$

$$\sigma_{abs}^{1841nm}(T) = -1.96x10^{-26} + 3.53x10^{-28}T \tag{23}$$

$$\sigma_{em}^{1841nm}(T) = 3.75x10^{-25} - 1.57x10^{-29}T \tag{24}$$

3. Results and discussion

As a first step, we analyze the power conversion along the fiber between the pump and signal radiations for the different taper schemes described in 0. In addition, we consider the following absorption and emission cross-sections for the Yb-doped fiber amplifier at $20^{\circ}C$: $\sigma^{p}_{abs} = 1.488x10^{-24}m^2$, $\sigma^{p}_{em} = 1.829x10^{-24}m^2$, $\sigma^{s}_{abs} = 6x10^{-27}m^2$, and $\sigma^{s}_{em} = 3.58x10^{-25}m^2$, where the wavelengths 976 and 1064 nm correspond to the pump and signal radiation, respectively. In the numerical simulation, we have fixed the numerical aperture of 0.18 along the tapered fiber, and we change the temperature from 20 to 120°C. In both taper structures named Taper 1 (**Figure 1a**) and Taper 2 (**Figure 1b**,) we employ a pump power of

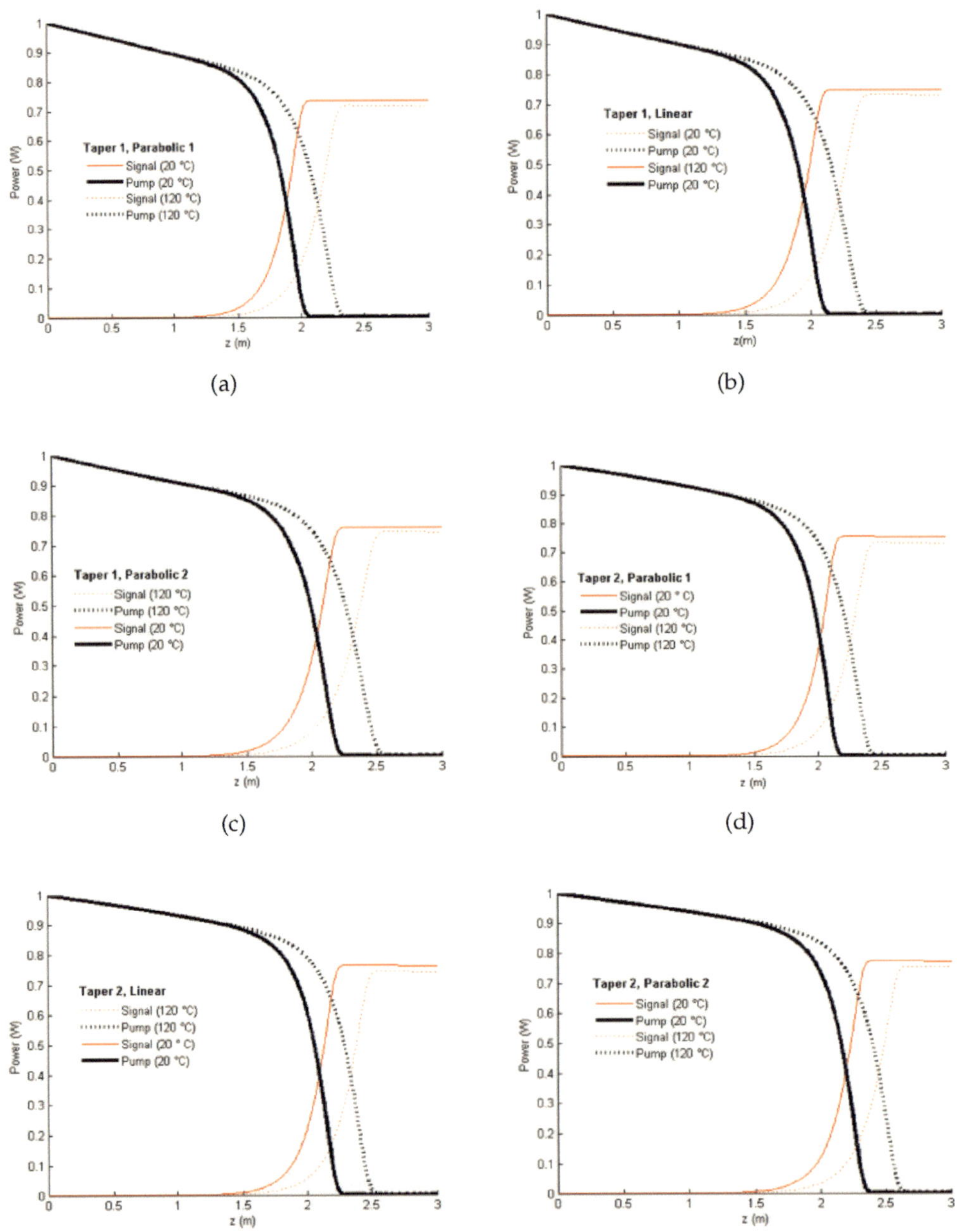

Figure 2.
Modeling of a tapered Yb-doped amplifier. Evolution of the pump and signal radiations at two different temperature;, (a), (b) and (c) for the scheme named taper 1 at 1 W of pump power for different tapered core shapes; (d), (e), and (f) for the scheme named taper 2 at 1 W of pump power for different tapered core shapes.

1 W, and the corresponding numerical results are shown in **Figure 2** for each tapered structure. In this figure it is relevant to observe the behavior of the cross-point where the signal and pump conversion occurs, due to the fact that this cross-point shifts to different lengths according to temperature variations. Then, this shift is an indicative of which tapered structure is affected principally by temperature, modifying in this way the Yb-doped fiber amplifier performance. To visualize this temperature response in a better way, we proceed to analyze the amplified signal power at the end of the tapered fiber at 20°C with respect to the signal generated at different temperatures T using the following normalized expression:

$$[P_s(20^\circ C) - P_s(T)]/P_s(20^\circ C). \tag{25}$$

We evaluate this equation for each tapered structure given in **Figure 1**. In the calculations we employ a fiber length = 3 m, and a pump power = 1 W. The results are shown in **Figure 3**. According to **Figure 3**, all tapered fiber structures show a high temperature sensitivity for large values of temperature. In addition, we can observe that the tapered fiber structure named Taper 2 is more sensitive to temperature for different tapered shapes. Especially, for the Taper 2 with a parabolic-1 shape, we obtain an improvement of the temperature sensitivity in the fiber amplifier.

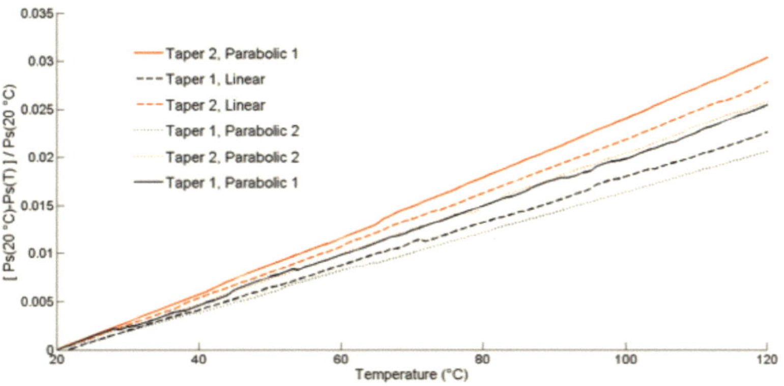

Figure 3.
Efficiency signal conversion for different temperatures at the end of the tapered fiber.

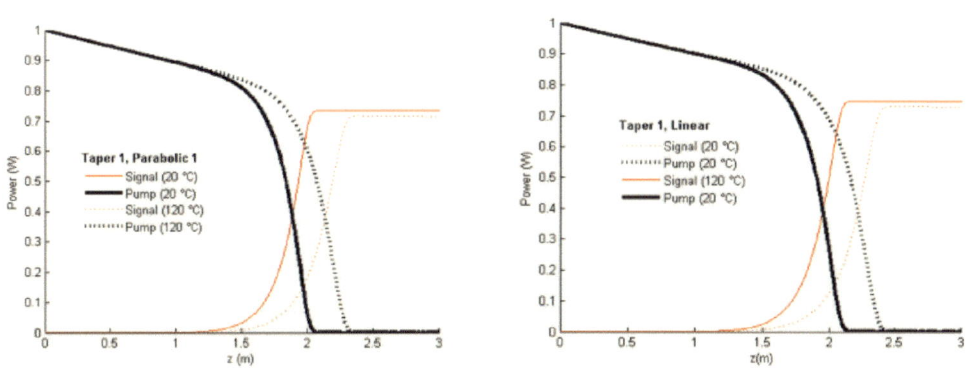

(a) (b)

Figure 4.
Temperature behavior of the generated signal for different temperatures T = 30 and 120°C at different pump powers: 0.1 − 0.5W.

Temperature response at different pump powers is shown in **Figure 4**, where temperatures of 30 and 120°C are used. The signal is calculated at the end of the tapered fiber (3 m) according to the scheme of **Figure 1**.

According to **Figure 4**, the temperature sensitivity of the signal radiation for both tapered fiber schemes shown in **Figure 1** is higher as the temperature is increased, and it can be improved if we employ lower pump powers than 1 W, respectively. In particular, this increment on sensitivity highly depends on the taper

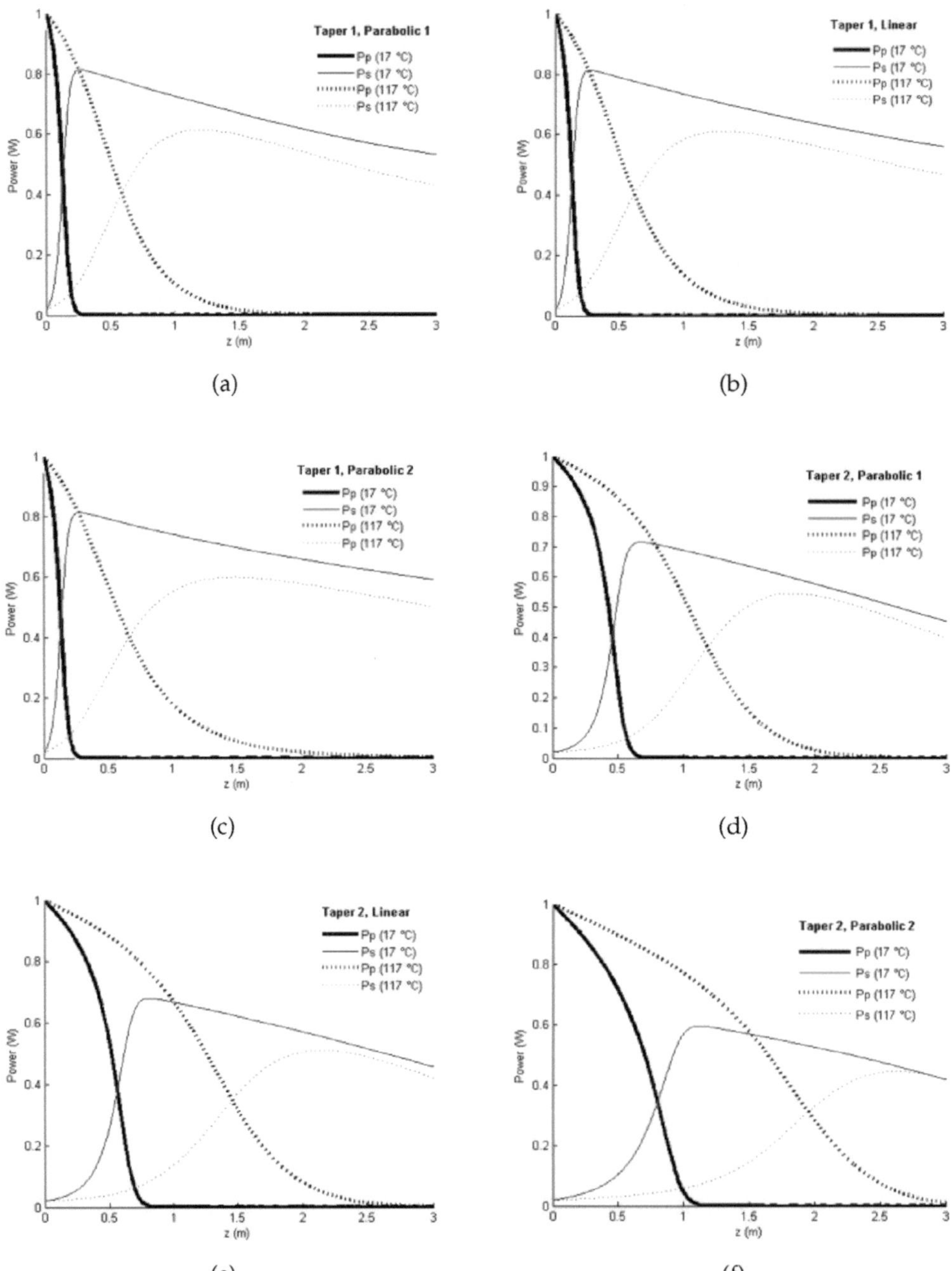

Figure 5.
Modeling of a tapered Tm-doped amplifier. Evolution of the pump and signal radiations at two different temperatures,; (a), (b), and (c) for the scheme named taper 1 at 1 W of pump power for different tapered core shapes; (d), (e), and (f) for the scheme named taper 2 at 1 W of pump power for different tapered core shapes.

shape and the pump scheme employed in the amplifier design. Additionally, this temperature sensitivity can change according to the tapered fiber length. In our calculations, the tapered length was 3 m, but other temperature sensitivities can be obtained if we consider tapered fiber length between 2 and 2.5 m, where the cross-point of the signal and pump radiations is located (See **Figure 2**). In this sense, further analysis needs to be performed to optimize the temperature sensitivity using different taper ratios and different taper lengths and modifying the longitudinal shape of the tapered doped fiber amplifier. Similarly, in the case of Th, we analyze the radiation conversion along the fiber between the pump and signal radiations of the different schemes described in **Figure 1**. The numerical aperture $NA = 0.18$ is considered constant along the taper, and the temperature is changed from 17–117°C. For both schemes shown in **Figure 1**, the co-propagating pump and signal power were set at 1 W and 20 mW, respectively; the corresponding results of the signal conversion for three different temperatures are shown in **Figure 5** for each longitudinal tapered core shape.

According to **Figure 6**, all tapered doped fiber amplifiers show an increasing temperature sensitivity for values starting even at 27°C and up to the maximum temperature of 117°C. The curves for "Taper 2" scheme (all in blue color) showed a saturation at around 87°C, indicating less sensitivity to temperature with respect to "Taper 1" scheme (in black color). However, this behavior could vary if we use other fiber lengths. In this context, the curves shown in **Figure 5** represent a design map that could allow us to choose the taper shape and its corresponding taper length, given by the cross-point in order to improve the temperature sensitivity of the tapered doped fiber amplifier around a specific temperature value.

On the other hand, **Figure 7** shows the temperature response at different pump powers for the two tapered core shapes analyzed in this work, where temperatures of 27 and 117°C are used. The signal is calculated at the end of the tapered fiber (3 m) according to the scheme of **Figure 1**.

According to **Figure 7**, the temperature sensitivity of the tapered doped fiber amplifier grows as the temperature is increased, and it is higher for low values of the pump power. This is an important result to consider for the design of fiber lasers and temperature fiber sensors. Therefore, we can consider that the temperature sensitivity of the signal radiation is higher for low pump powers, and this sensitivity is different for each pump scheme and depends on the taper used. This sensitivity can vary according to the length of the fiber used. For example, these calculations were made at 3 m; however, more sensitivity changes can be obtained at shorter length, as it is shown in **Figure 2**. Additionally, further analysis needs to be

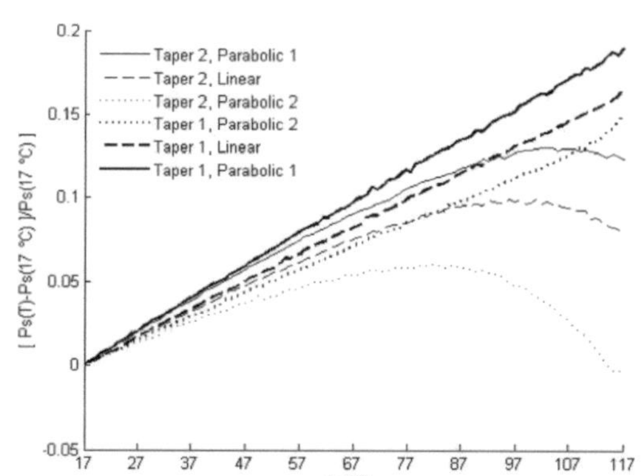

Figure 6.
Normalized amplified signal for different temperatures at the end of the tapered fiber with L = 3m.

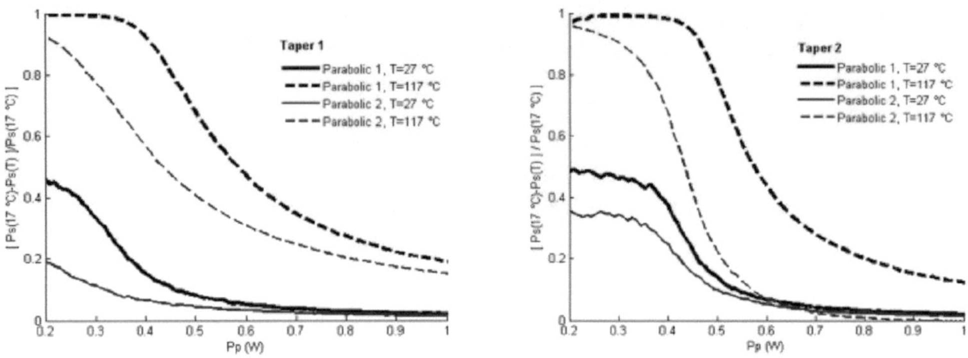

(a) (b)

Figure 7.
Temperature behavior of the generated signal for temperatures T = 300 and 390°K at different pump powers (0.2–1 W): (a) taper 1, parabolic 1 and 2, (b) taper 2, parabolic 1 and 2.

Sensing material	λ pump (nm)	Fluorescence wavelengths (nm)	Temperature range (°C)	Sensitivity (°C^{-1})	Reference
Er: silica fiber	800	530/555	23–600	0.013	[10]
Yb: silica fiber	810	910/1030	20–600	0.095	[31]
Nd: silica fiber	807 nm	820–840/895–815	−50–500	0.168	[3]
Er: silica fiber	785	527–537/545–555	21–96	0.006	[8]
Yb: tapered silica fiber	976	1064	15–120	0.005	This work
Tm: tapered silica fiber	1650	1841	17–117	0.009	This work

Table 1.
Comparison of the performance of rare-earth-doped fibers and materials as temperature-sensing elements.

performed to optimize the temperature sensitivity using different taper ratios and the longitudinal shape of the tapered doped fiber.

Similar results were obtained as in the case of using an Yb-doped fiber amplifier for temperature sensing. However, although the response follows a similar behavior, the results are different, the evolution of the pump and signal radiation is shifted to shorter lengths along the fiber, the efficiency signal conversion is larger, and the temperature behavior of the generated signal for different temperatures as a function of the pump power shows an increased response for the Tm-doped fiber amplifier. The sensitivity of both tapers can be compared with other kinds of sensor of temperatures using similar techniques as shown in **Table 1**. **Table 1** compares the parameters of the authors' tapered ytterbium- and thulium-doped fiber temperature sensors with existing ones. It can be seen that the sensitivity of the sensors is close to early work on doped silica fibers for intrinsic fiber-optic temperature sensors.

4. Conclusion

We have reported a numerical analysis of the temperature effects in tapered ytterbium and thulium-doped fiber amplifiers in co-propagation mode. We found that the tapering shape and the pump power at the input amplifier modify significantly its temperature sensitivity. In particular, the temperature sensitivity is higher as the core radius is reduced in the propagation direction (Taper 1 scheme), and it can be maximized if the tapered fiber length is equal to the propagation distance where the cross-point of the power conversion is achieved. On the other hand, the temperature sensitivity can also be incremented if we use lower pump powers at the amplifier input. Then, for temperature sensing applications, it is desirable to work using tapered doped fiber amplifiers in low pump power regimes. Our results can be extrapolated to other doped fibers and are of great interest for the improvement of high-power tapered lasers and the development of temperature fiber laser sensors.

Acknowledgements

The authors gratefully acknowledge UNACAR and the Optical Sciences Group at The University of Twente.

Conflict of interest

The authors declare no conflict of interest.

Author details

Rafael Sanchez-Lara[1*†], Lelio de la Cruz-May[1†], José Luis Vazquez-Avila[1†], Daniel Enrique Ceballos-Herrera[2†] and José Alfredo Alvarez-Chavez[3†]

1 Facultad de Ingenieria, Universidad Autonoma del Carmen (UNACAR), Ciudad Del Carmen, Campeche, Mexico

2 Instituto de Ingenieria, Universidad Nacional Autonoma de Mexico (UNAM), Cd. Universitaria, Mexico

3 Optical Sciences Group, University of Twente, Enschede, The Netherlands

*Address all correspondence to: rafaunacar@gmail.com

† These authors contributed equally.

References

[1] Sukkar K, Mahmood S, Alhuda H, Jabbar N. Temperature sensing for petroleum pipelines monitoring and control using fiber optics. In: 3rd International Scientific Conference New Trends in Information Technology and Communications Applications (NTICT). Baghdad: In Cooperation with IEEE and UOITC; 2017

[2] Amini S, Kavousi P, Carr TR. Application of fiber-optic temperature data analysis in hydraulic fracturing evaluation: A case study in marcellus shale. In: Unconventional Resources Technology Conference (URTeC'2017). Austin, Texas: Society of Exploration Geophysicists, American Association of Petroleum; 24-26 July 2017. pp. 2105-2115

[3] Wade S, Baxter G, Collins S, Grattan K, Sun T. Simultaneous strain–temperature measurement using fluorescence from Yb-doped silica fiber. The Review of Scientific Instruments. 2000;**71**(6):2267-2269

[4] Liu T, Fernando G, Zhang Z, Grattan K. Simultaneous strain and temperature measurements in composites using extrinsic Fabry–Perot interferometric and intrinsic rare-earth doped fiber sensors. Sensors and Actuators, A: Physical. 2000;**80**(3): 208-215

[5] Trpkovski S, Wade SA, Collins SF, Baxter GW. Er3+:Yb3+ doped fibre with embedded FBG for simultaneous measurement of temperature and longitudinal strain. Measurement Science and Technology. 2005; **16**(2):488

[6] Jung J, Nam H, Lee JH, Park N, Lee B. Simultaneous measurement of strain and temperature by use of a single-fiber Bragg grating and an erbium-doped fiber amplifier. Applied Optics. 1999;**38**(13):2749-2751

[7] Wade SA, Collins SF, Baxter GW. Fluorescence intensity ratio technique for optical fiber point temperature sensing. Journal of Applied Physics. 2003;**94**(8):4743-4756

[8] Paez G, Strojnik M. Experimental results of ratio-based erbium-doped-silica temperature sensor. Optical Engineering. 2003;**42**(6):1805-1812

[9] Maurice E, Monnom G, Dussardier B, Saïssy A, Ostrowsky D, Baxter G. Erbium-doped silica fibers for intrinsic fiber-optic temperature sensors. Applied Optics. 1995;**34**(34): 8019-8025

[10] Castrellon-Uribe J, García-Torales G. Remote temperature sensor based on the up-conversion fluorescence power ratio of an erbium-doped silica fiber pumped at 975 nm. Fiber and Integrated Optics. 2010;**29**(4):272-283

[11] Zhu S, Xing JW, Pang FF, Wang TY. Temperature sensor based on a single-mode tapered optical fiber. Journal of Shanghai University (English Edition). 2011;**15**(2):101

[12] Kim KT, Park KH. Fiber-optic temperature sensor based on single mode fused fiber coupler. Journal of the Optical Society of Korea. 2008;**12**(3): 152-156

[13] Prerana P, Varshney RK, Pal BP, Nagaraju B. High sensitive fiber optic temperature sensor based on a side-polished single-mode fiber coupled to a tapered multimode overlay waveguide. Journal of the Optical Society of Korea. 2010;**14**(4):337-341

[14] Díaz-Herrera N, Navarrete M, Esteban O, González-Cano A. A fibre-optic temperature sensor based on the deposition of a thermochromic material

on an adiabatic taper. Measurement Science and Technology. 2003;**15**(2):353

[15] Villatoro J, Monzón-Hernández D, Mejía E. Fabrication and modeling of uniform-waist single-mode tapered optical fiber sensors. Applied Optics. 2003;**42**(13):2278-2283

[16] Zhu S, Pang F, Wang T. Single-mode tapered optical fiber for temperature sensor based on multimode interference. In: Asia Communications and Photonics Conference and Exhibition (ACP'2011). Shanghai, China: Optical Society of America; 13-16 November 2011. p. 83112B

[17] Monzón-Hernández D, Mora J, Perez-Millán P, Díez A, Cruz JL, Andrés MV. Temperature sensor based on the power reflected by a Bragg grating in a tapered fiber. Applied Optics. 2004;**43**(12):2393-2396

[18] Henry LJ, Shay TM, Hult DW, Rowland KB. Thermal effects in narrow linewidth single and two tone fiber lasers. Optics Express. 2011;**19**(7): 6164-6176

[19] Vazquez-Zuniga LA, Chung S, Jeong Y. Thermal characteristics of an ytterbium-doped fiber amplifier operating at 1060 and 1080 nm. Japanese Journal of Applied Physics. 2010;**49**(2R):022502

[20] Peng X, Dong L. Temperature dependence of ytterbium-doped fiber amplifiers. JOSA B. 2008;**25**(1):126-130

[21] Newell T, Peterson P, Gavrielides A, Sharma M. Temperature effects on the emission properties of Yb-doped optical fibers. Optics Communication. 2007; **273**(1):256-259

[22] Saito K, Yamamoto R, Kamiya N, Sekiya E, Barua P. Fictive temperature dependences of optical properties in Yb-doped silica glass. In: Proceeding of Spie 2008. San Diego, California, United States: International Society for Optics and Photonics; 10-14 August 2008. p. 69981-J

[23] Filippov V, Chamorovskii Y, Kerttula J, Golant K, Pessa M, Okhotnikov O. Double clad tapered fiber for high power applications. Optics Express. 2008;**16**(3):1929-1944

[24] Li L, Lou Q, Zhou J, Dong J, Wei Y, Du S, et al. High power single transverse mode operation of a tapered large-mode-area fiber laser. Optics Communication. 2008;**281**(4):655-657

[25] Alvarez-Chavez J, Grudinin A, Nilsson J, Turner P, Clarkson W. Mode selection in high power cladding pumped fibre lasers with tapered section. In: Technical Digest. Summaries of papers presented at the Conference on Lasers and Electro-Optics (CLEO'99). Baltimore, MD, USA: Conference on Lasers and Electro-Optics IEEE; 28-28 May 1999. p. 247-248

[26] Ustimchik V, Nikitov S, Chamorovskii YK. Simulation of radiation generation in an active double-clad optical tapered fiber. Journal of Communications Technology and Electronics. 2011;**56**(10):1249

[27] Bagan V, Nikitov S, Chamorovskii YK, Shatrov A. Studying the properties of double-clad active cone optic fibers. Journal of Communications Technology and Electronics. 2010;**55**(10):1154-1161

[28] Zhang YJ, Zhong FF, Wang YZ. Top-hat beam Tm3+ −doped fiber laser using an intracavity abrupt taper. Laser Physics. 2011 Jan;**21**(1):215-218. DOI: 10.1134/S1054660X11010245

[29] Ghatak A, Thyagarajan K. An introduction to fiber optics. 1st ed. Cambridge: Cambridge University

Press; 1998. p 576. DOI: 10.1017/CBO9781139174770

[30] Alvarez-Chavez JA, Perez-Sanchez G, Ceballos-Herrera D, Rodriguez-Rodriguez J, Schreiber T. Temperature sensing characteristics of tapered Yb-doped fiber amplifiers. Optik. 2013; **124**(22):5818-5821

[31] Sanchez-Lara R, Ceballos-Herrera D, Vazquez-Avila J, de la Cruz-May L, Martinez-Pinon F, Alvarez-Chavez J. Temperature sensing characteristics of tapered Tm3+—doped fiber amplifiers. Laser Physics. 2017;**27**(8):085108

[32] Turri G, Sudesh V, Richardson M, Bass M, Toncelli A, Tonelli M. Temperature-dependent spectroscopic properties of Tm3+ in germanate, silica, and phosphate glasses: A comparative study. Journal of Applied Physics. 2008; **103**(9):093104

[33] Sudesh V, Piper J, Goldys E, Seymour R. Growth, characterization, and laser potential of Tm: $La_2Be_2O_5$. JOSA B. 1998;**15**(1):239-246

[34] Agger SD, Povlsen JH. Emission and absorption cross section of thulium doped silica fibers. Optics Express. 2006;**14**(1):50-57

[35] Payne SA, Chase L, Smith LK, Kway WL, Krupke WF. Infrared cross-section measurements for crystals doped with Er3+, Tm3+, and Ho3+. IEEE Journal of Quantum Electronics. 1992;**28**(11):2619-2630

Chapter 7

Fiber-Optic Temperature Sensors with Chalcogenide Glass and Crystalline Sensing Element

Igor Chychura

Abstract

Fiber-optic temperature sensors (FOTSs) are competitive in cases where it is necessary to provide stable operation in harsh operating conditions—strong electromagnetic fields, elevated background radiation, and explosive environment—and at large distances between sensing element and receiving device. In addition to high reliability and stability when measurements are performed in extreme conditions, such sensors have good metrological characteristics. For FOTSs the main functional node is a temperature-sensitive element (TSE), which defines the main operational and metrological characteristics of these devices. Therefore, the search for new promising materials for innovative TSEs is relevant. From this point of view, chalcogenide amorphous materials of the As_xSe_{100-x} system deserve special attention. They are characterized by a unique combination of high inertia under the influence of intensive external factors and the possibility of significant changes of their optical parameters with changing of their chemical composition within the range of glass formation. In this scientific work, a comparative analysis of opportunities for traditional GaAs crystals and As_xSe_{100-x} chalcogenide glasses for TSEs of modern FOTSs is carried out.

Keywords: fiber-optic temperature sensor, temperature-sensitive element, transfer characteristics, chalcogenide glass semiconductor materials, optical transmission

1. Introduction

Nowadays, it is impossible to imagine modern industry without automated control systems. Complex structures and mechanisms require constant control of various parameters and processes that happen in them. Monitoring of complex engineering structures and industrial systems is an integral part of their day-to-day operation. To perform such control, sensors and control systems of various physical quantities such as temperature, mechanical deformation, pressure, etc. are required.

At present, the market of measuring systems and sensors was dominated by electronic measuring technologies, which involve the transformation of the measuring parameter into an electrical signal and its subsequent processing [1]. An alternative to such an approach is the use of fiber-optic measurement systems, where the measured parameter is converted into an optical signal transmitted over an optic fiber. The market of fiber-optic sensors has been growing from the moment they appeared, 2.5 billion dollars in 2004 with an annual increase of 11% in all industries, according to Frost and Sullivan's marketing agency. Fiber-optic sensors have some advantages over more traditional electronic technologies: (a) explosion safety; (b) zero sensitivity to electromagnetic interference (noise immunity); and (c) high separation ability. Despite all mentioned above, the relative share of fiber-optic sensors in the overall market of measuring systems remains small. In essence, fiber-optic sensors occupy only a very small position where traditional measuring instruments cannot be used or their use is very expensive.

However, it is worth noting two trends that are observed in our time. Firstly, the rapid development of related technologies, fiber-optic transmission of information, reception and processing of images using digital photo and video equipment, microprocessor technology, contributes to the development of fiber-optic measuring technology and in cheapening their production. Secondly, industry and regulators put increasingly stringent requirements on the operating conditions of such devices, namely, noise immunity requirements, measurement safety, accuracy, etc. These are the criteria that can satisfy modern fiber-optic sensors. It can be seen that these two trends can lead to a situation where the fiber-optic measuring systems will become more competitive in than traditional systems [2]. This is due, not only, to the performance characteristics but also to the cost of individual measuring channel.

2. Types of fiber-optic sensors

Fiber-optic sensors can be classified by many parameters, one of which is the classification based on the coding of measured information:

- Phase sensors using a highly coherent source of radiation and measuring the phase of a light wave that changes under the influence of the external parameters [3].
- Spectral coding sensors, which, in contrast to the purely phase, use radiation sources with a wide spectrum and the possibility of analyzing the entire spectrum [4].
- Amplitude sensors, in which the measured parameters modulate the intensity of light waves that passed through the sensor or are reflected from it [5].
- Polarizing sensors, using information about the polarization of the light wave [6].

Another classification of fiber-optic sensors is by the work principle:

- Interference [7].
- Distributed [8].
- Luminescent [9].
- Using the gratings inside optic fibers [10].
- Combined [11].

You can also classify sensors by localization of the measured parameter:

- Pointing.
- Distributed.
- Imaginary distributed.

All of the mentioned fiber-optic sensors have certain advantages and disadvantages associated with their design, operation principle, etc. All of this affects on their price and application possibilities.

For example, phase sensors using laser radiation sources are quite common but mostly in lab installations and not in industry. This is due to the needs of accurate adjustment of devices and presence of additional phase-adjustment schemes, which greatly complicate their design. In addition, such sensors do not allow measurements of absolute values. To eliminate these disadvantages, several frequencies of optic radiation are used, which makes this method an intermediate type of measurements between phase and spectral [12], which are presented below.

Sensors with spectral coding are mostly promising in terms of their implementation into the industry due to their resistance to various parasitic parameters: drift of radiation power source, uncontrolled power loss in fibers, losses due to fiber coupling using connectors, etc. In addition, this type of sensors allows measurements of absolute values and does not require recalibration after switching off the instrument. For now, this measurement method was considered very difficult and expensive. It needed the presence of a spectrometer and optic image processing equipment. But the situation is changing, and due to the cheapening of the methods of optic spectrum processing, development of microprocessor technology and the technology of receiving an optic image, and cheapening of optoelectronic components, the price of the fiber-optic measurement channel comes close to electronic analogues [13].

Amplitude sensors have their area of application due to their low cost and can be used where there is no need for high measurement accuracy (e.g., as counters of rotation, microphones, temperature distribution sensors, etc.). However, in high-precision measuring systems, they are not widely used due to their relatively low accuracy and probability of parameter drift.

Tunnel sensors are highly sensitive devices, but they also have parameter drift, so they can only find limited use, for example, in high-precision positioning devices, microphones, hydrophones, etc.

Polarizing sensors, in essence, are analogues to interference sensors. Their commercialization is mainly hampered by the need to use expensive fiber while maintaining polarization.

3. Fiber-optic temperature sensors with amplitude modulation of light

From the types of fiber-optic sensors being represented in the previous paragraph, sensors with amplitude modulation of light are most suitable for further processing of the output signal from optical fiber. Most of the schemes with amplitude modulation of signals do not require the use of coherent light and therefore do not include specific requirements for emission source (LED) and emission detector (photodiode).

The amplitude modulation of optic signal can be done in several ways: (1) attenuation of light when α absorption coefficient changes; (2) changes in the cross section on the way of light passing; (3) changes in the coefficient of reflection of the medium when refractive index n is changing; (4) receiving additional radiation when the temperature of the sensitive element of sensor is changing; and others.

The operation principle of the first-type fiber-optic sensors is that light passing through a semiconductor material is absorbed in the case when the photon energy $h\nu$ is greater than the width of the bandgap *Eg* [14].

With increasing temperature, the width of the bandgap decreases monotonously. In works [5, 15–17], as a temperature sensor, a *GaAs* crystal was selected. The principle of operation of the sensor is to shift the edge of optic absorption induced by temperature change (**Figure 1**). From the graph, it is seen that, for example, passing the sample at wavelength λr will be different when temperature changes. Knowing the optic characteristics of a material, one can estimate the transmission characteristic of such a sensor.

Indeed, according to the Beer–Lambert law of absorption

$$I(l, T) = I_0(1 - R)\exp\left[\alpha(T) \cdot l\right] \quad (1)$$

where:

I_0 is the intensity of the falling light.

I is the intensity of the light passing through the sample.

R is coefficient of reflection.

$\alpha(T)$ is the material absorption coefficient at a given temperature and wavelength.

As shown in [16], for *GaAs*.

$$\alpha(T) = A \cdot \left[h\nu - E_g(T)\right]^{1/2}, \quad (2)$$

where:

A is the constant of the material.

$E_g(T)$ is the width of the bandgap at temperature T.

h is Planck's constant.

ν is the light frequency.

In the temperature range of 20–297 K, $E_g(T)$ can be described by the formula:

$$E_g(T) = \frac{E_g(0) - \gamma T^2}{(\beta + T)}, \quad (3)$$

where $E_g(0)$ is the width of the forbidden zone at 0 K and γ and β are empirical constants.

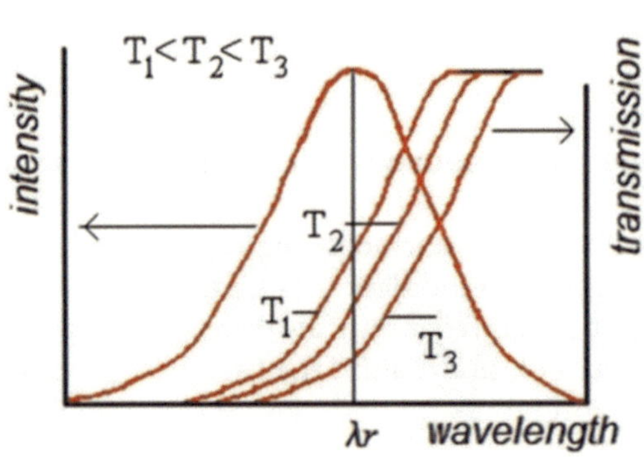

Figure 1.
The principle of operation of the fiber-optic temperature sensor: (1) spectrum of light source radiation; (2, 3, 4) the dependence of the optic transmittance of the material on the wavelength λ at different temperatures T2 < T3 < T4.

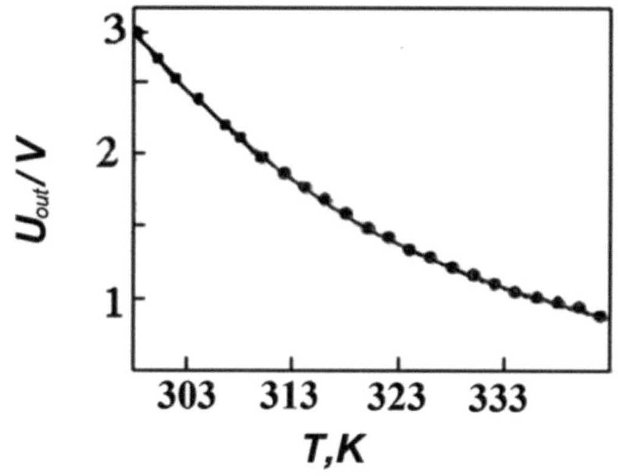

Figure 2.
Dependence of the photodiode output voltage from temperature sensor temperature [2].

For *GaAs*, $E_g(0)$ = 1.522 *eV*, γ = 5.8 · 10-4E/K, β = 300 K, A = 2.446 · 10-4 (cm · eV)$^{-1}$. Proceeding from Eq. (3), we can find the wavelength λ_g of light on which a given semiconductor will absorb:

$$\lambda_g(T) = \frac{n_c}{E_g(T)} = \frac{n_c}{\frac{E_g(0) - \gamma T^2}{(\beta + T)}} \tag{4}$$

Taking into account the Eqs. (2) and (3), the connection between the intensity of light I passing through the sensitive element and its temperature can be written as

$$I(l, T) = I_0(1 - R) \exp\left\{A\left[h\nu - E_g(0) + \frac{\gamma T^2}{T + \beta}\right]^{½} \cdot l\right\} \tag{5}$$

Thus, if the light passes through the sample in the region of its absorption, the intensity of light will diminish, with increasing of sample temperature. If the emission detector (photodiode) is chosen in the working range, then change in intensity of light will lead to a decrease in the photodiode voltage (**Figure 2**).

4. Fiber-optic temperature sensor with chalcogenidic glass sensitive element

The operation principle of such fiber-optic sensors is similar to the one discussed above. Change in the light transmission in chalcogenide vitreous semiconductor (ChVS) with temperature change. The use of as sensitive elements of ChVS gives the possibility for better optic coordination of emission source (LED), temperature-sensitive element, and emission detector (photodiode), since the physical and chemical, optic properties of glass can be changed in wide range by changing their composition within the region of glass formation. Developed FOTS are using a film of single crystals *GaAs* and *Ge*, with given optic constants, as a sensitive element. Our previous studies [18] have shown that, as an active element of FOTS, from a large number of ChVS can be used *As-Se* glass system.

In few scientific and technical publications, several analogues of FOS with semiconductor crystals as a temperature-sensitive element were published [5, 15–17]. Often, these are two-wave single-channel fiber-optic circuits, which use two LEDs

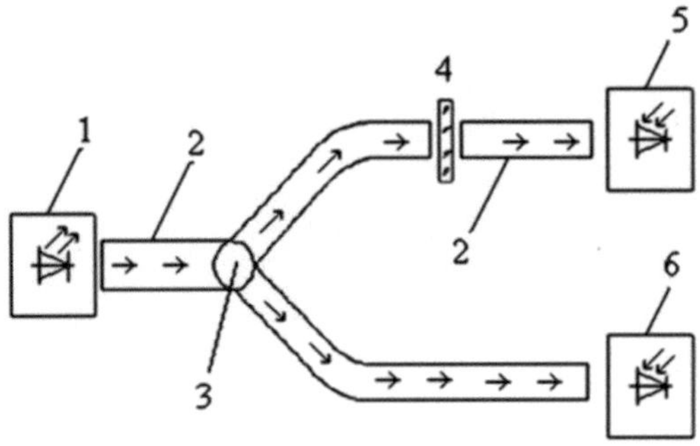

Figure 3.
Optic scheme of the primary converter: 1, radiation source (LED); 2, fiber-optic coupler; 3, place of optic signal division; 4, temperature-sensitive element; 5, the first photodiode (FD1); and 6, the second photodiode (FD2).

with different wavelengths (operating and reference). If the ambient temperature changes, the change in characteristics will not be the same, and the compensation scheme will not be able to eliminate the measurement error in this case. In our view, a more stable scheme of temperature measurement with a fiber-optic sensor is a single-LED dual-channel circuit, which is shown in **Figure 3**.

The operation principle of this type of FOS is that light signal after passing through the Y-cutter on two channels (working and reference) is received by two identical photodiodes. The registration system measures the ratio of signals from photodiodes 5 and 6 (**Figure 3**). Therefore, changes that can occur in the fiber-optic line for a given temperature will not affect the ratio of signals on photodiodes.

In scientific work (article) [18], we developed a method for performing calculations of the transmission characteristics of the fiber-optic temperature sensor. In this chapter, a method of selecting the elements of the primary measuring transformer and the optic parameters of LED, photodiode, and sensitive element of such fiber-optic temperature sensor was simulated. The results of the transfer characteristic in the form of a plot of the dependence of the change in the voltage of the emission detector (photodiode) on the temperature of the medium in which the temperature-sensitive element is shown demonstrate the performance of this technique.

In this chapter, a plate of ChVS *As-Se* was used as a temperature-sensitive element. The change in the composition of the components of the temperature-sensitive element allowed us to ensure the optimum harmonization of the spectral characteristics of the receiver source and the temperature-sensitive element, which are very convenient for the construction of fiber-optic thermometers. However, not only ChVS can be used as a temperature-sensitive element in a fiber-optic temperature sensor; some crystalline materials can also be effectively implemented in these devices.

5. Fiber-optic temperature sensor with crystalline sensitive element

Preliminary analysis of the probable variants of crystalline materials which can be used as temperature-sensitive elements indicated the possibility of using base materials of the A_3B_5 (*GaAs*, *GaP*), which today are the basic semiconductor crystals for modern microelectronics [19, 20].

From the practical point of view, the interest in these materials is due to their wide application in the elements of infrared optics, television equipment, and fiber-optic communication. These materials are resistant to radiation, which makes their practical application under conditions of radioactive exposure possible, and the possibility of anatomy of radioactive defects opens the possibility of their multiple use in extreme radiation conditions. In addition, the optical properties of these crystals are immune to high-frequency electromagnetic fields.

All of this indicates a possibility of effective application of *A3B5* crystals as sensitive elements for amplitude-type FOS for temperature, pressure, and other physical quantities.

For modeling parameters of primary measuring converter, *GaP* crystals doped with *Zn* (*GaP:Zn*) were used. These crystals are used to construct different types LEDs [21].

For experimental studies of optical transmittance, special samples were made. The methods of making are as follows: crystal plates were cut out of solid material and polished on abrasive powders, then polished on diamond paste of different consistency, and electropolished to a high class of roughness. As a result, we got plates with a thickness of 300 microns and high-quality surface.

Figure 4 shows a picture of one of the crystal *GaP:Zn*. This plate was used in our optical studies.

Results of these studies, in the form of direct records, were made on a specialized optical installation.

According to the transmission spectrum, which is shown in **Figure 5**, the optical absorption spectra at different temperatures, shown in **Figure 6**, were calculated.

Optical absorption in *GaP:Zn* crystals according to the literature data [22] is indirect banded, which determines the angle of absorption dependence from wavelength. For the convenience of determining the width of the bandgap at different temperatures, the dependence $\alpha_2 = f\,(hv)$ was constructed (**Figure 7**).

Subsequent processing on the computer obtained experimental results in the form of transmission spectra of *GaP:Zn* crystal at various temperatures which is shown in **Figure 8**. It can be seen from **Figure 8** that with increasing of temperature, the transmittance dependence shifts to greater wavelengths and overall bandwidth drops. This allowed us to determine the wavelength position which will determine the controlled level of transmission at given temperature. It was most convenient to use a λ = 600 nm, which was determined by the level of transmission at a fixed temperature. The results of the analysis of the temperature shift of the transmission level, determined by the spectra in **Figure 8**, are presented in **Figure 9**.

Dependence of the absorption coefficient α in non-triangular semiconductors to which *GaP* belongs is described by the dependence:

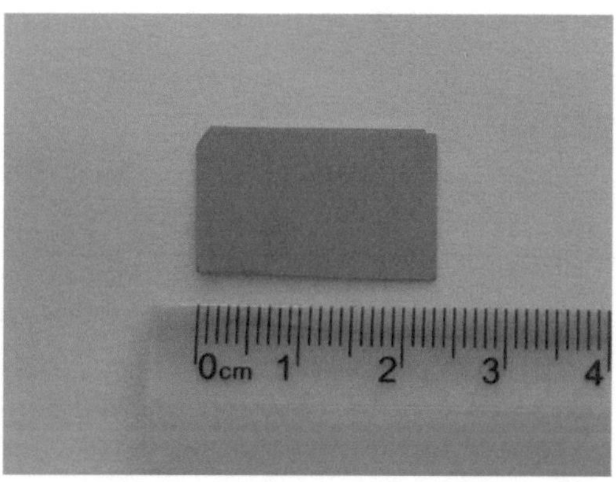

Figure 4.
Photograph of GaP:Zn *crystal used in optical studies.*

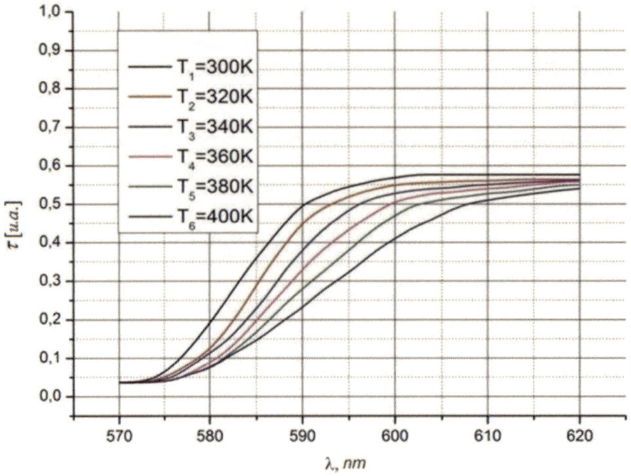

Figure 5.
GaP:Zn crystal spectrum at different temperatures T (K).

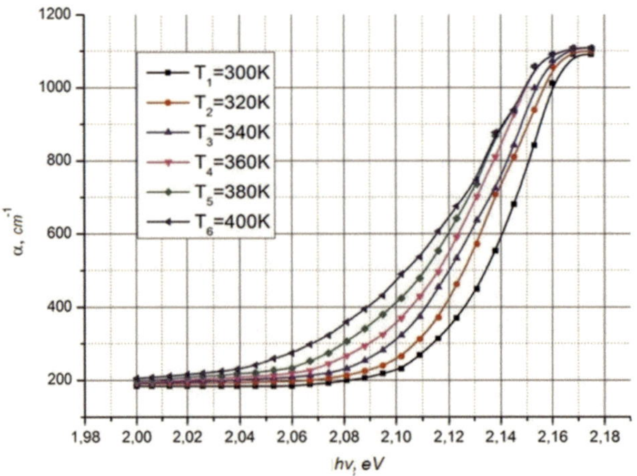

Figure 6.
Dependence of absorption of GaP:Zn *crystals at different temperatures T (K).*

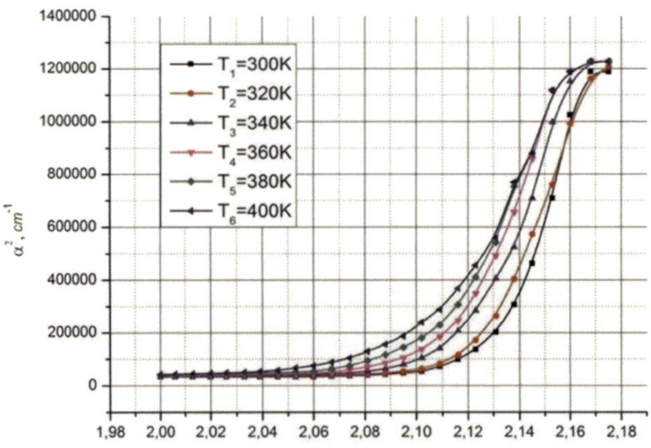

Figure 7.
Dependence of absorption of GaP:Zn *crystal at different temperatures T (K).*

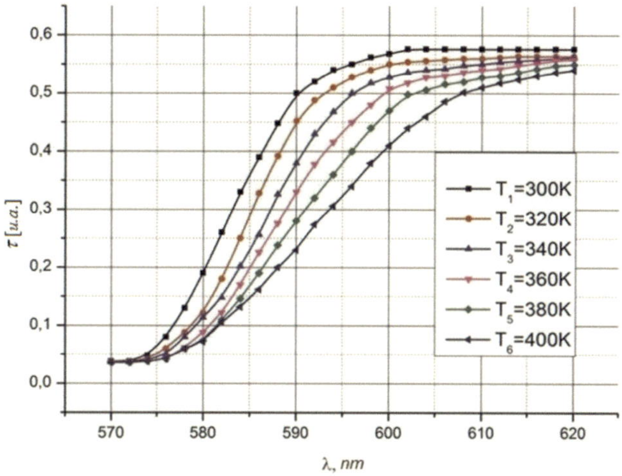

Figure 8.
Spectrum of GaP:Zn *crystal at different temperatures T (K).*

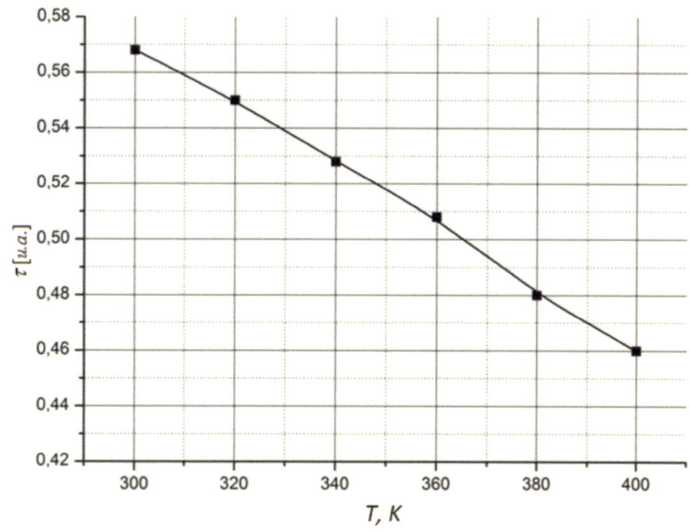

Figure 9.
Temperature dependence of transmission τ_λ on $\lambda\tau$ = 600 nm for the GaP:Zn crystal.

$$\alpha = \alpha_0 \left[hv - E_g\right]^{\frac{1}{2}} \quad (6)$$

This means that when α_2 dependence is built, it is possible to obtain the expression:

$$\alpha^2 = {\alpha_0}^2 \left[hv - E_g\right] \quad (7)$$

From which it is easy to obtain the value of the width of the indirect bandgap *Eg* by extrapolating to the zero of the linear part of the dependence $\alpha^2 = f\ (hv)$.

Analyzing **Figure 8**, we can select the required operating wavelength to demonstrate the practical use of GaP:Zn crystals as a temperature-sensitive element of a fiber-optic temperature sensor. As a result we obtained temperature dependence of transmission of GaP:Zn crystal (**Figure 9**).

It can be seen from **Figure 9** that the passage of the *GaP:Zn* plate at a controlled wavelength λ = 600 nm linearly decreases in the temperature range from 300 to 400 K. This, not directly, indicate that the expected and calculated full-cycle photocurrent will also change with temperature change. That suggests the possibility of using GaP:Zn crystals as temperature-sensitive element of a fiber-optic temperature sensor.

6. Conclusions

Currently, fiber-optic sensors do not occupy a leading position in the measurement system market. However, the continuous development of technologies, including fiber-optics, makes such devices more competitive and provides consumers with the advantages that traditional measurement systems cannot give. This, in turn, expands their scope. In this paper, our focus was on fiber-optic temperature sensors, namely, sensors with amplitude modulation of light. These sensors are just one of a wide variety of modern fiber-optic devices. In our previous work, which was briefly presented in Section 4, we used ChVS materials as a temperature-sensitive element for a fiber-optic temperature sensor. Further studies have shown the possibility of using A3B5 crystals as a temperature-sensitive element also. Section 5 presents the possibility of using such semiconductor materials on the example of GaP:Zn crystals. The temperature dependence shown in **Figure 9** confirms that A3B5-type semiconductor materials can be used as a temperature-sensitive element of a fiber-optic sensor.

Author details

Igor Chychura
Engineering Department, National Universiry of Uzhhorod, Uzhhorod, Ukraine

*Address all correspondence to: igor.chechura@uzhnu.edu.ua

References

[1] Webster JG. The Measurement, Instrumentation and Sensors Handbook (Electrical Engineering Handbook). 1st ed. DOI: 10.1201/9780415876179.sec1

[2] Sabri N, Aljunid S, Salim M, Fouad S. Fiber Optic Sensors: Short Review and Applications; 2015. DOI: 10.1007/978-981-287-128-2_19

[3] MacDougall TW, Mendez A, Krohn DA. Fiber Optic Sensors: Fundamentals and Applications. 4th ed. 2015. DOI: 10.1117/3.1002910.ch4

[4] Pilipovich VA, Zalesskiy VB, Konoiko AI, Goncharenko I, Basalaev SP. Optical Fiber Linear Displacement Sensor with Spectral Coding on the Base of Fabry-Perot Interferometer; 2013

[5] Ding Y, Dai X, Zhang T. Low cost fiber-optic temperature measurement system for high voltage electrical power equipment. IEEE Transactions on Instrumentation and Measurement. 2010;**59**(4):923-933

[6] Woliński TR. Polarization in optical fibers. Acta Physica Polonica. 1999;**95**(5):749-760. DOI: 10.12693/aphyspola.95.749

[7] Noor SFSM et al. Multimode interference based fiber-optic sensor for temperature measurement. Journal of Physics: Conference Series. 2019;**1151**: 012023. DOI: 10.1088/1742-6596/1151/1/012023

[8] Xiaoyi B, Liang C. Recent progress in distributed fiber optic sensors. Sensors (Basel, Switzerland). 2012;**12**: 8601-8639. DOI: 10.3390/s120708601

[9] Grattan KTV, Zhang ZY, Sun T. Luminescent optical fibers in sensing. In: Grattan KTV, Meggitt BT, editors. Optical Fiber Sensor Technology. Optoelectronics, Imaging and Sensing. Vol. 4. Dordrecht: Springer; 1998. DOI: 10.1007/978-94-017-2484-5_8

[10] Eisenmann T. Fiber optical sensing with fiber bragg gratings. In: Münch U, Buchmann A, editors. GEOTECHNOLOGIEN Science Report. Vol. 15. Fürstenrieder Straße, München: NFAP GmbH; 2010. pp. 79-86. DOI: 10.2312/GFZ.gt.15.08

[11] Pevec S, Donlagić D. Multiparameter fiber-optic sensors: A review. Optical Engineering. 2019;**58**(7): 072009. DOI: 10.1117/1.OE.58.7.072009

[12] Liao Y, Austin E, Nash PJ. Phase Sensitivity Characterization in Fiber-Optic Sensor Systems Using Amplifiers and TDM

[13] Gudkov YI, Zak EA. Use of microprocessors in devices utilizing fiber-optic sensors. Measurement Techniques. 1996;**39**:1196. DOI: 10.1007/BF02375404

[14] Pankove JI. Optical Processes in Semiconductors (Dover Books on Physics) Paperback; 2010

[15] Soloman S. Temperature sensor in process control. Industrial sensor and control–The basics-part II; 2011

[16] Zhao Y, Rong M, Liao Y. Fiber-optic temperature sensor used for oil well based on semiconductor optical absorption. IEEE Sensor Journal. 2003; **3**(4):400-403. DOI: 10.1109/JSEN.2003.815790

[17] Lei XM. The study of the dual-wavelength Fiber-optic temperature sensor. Applied Mechanics and Materials. 2011;**128-129**:741-744. DOI: 10.4028/www.scientific.net/AMM.128-129.741

[18] Chychura II, Turianytsia II, Kozusenok OV. Transmission characteristic of fiber optic temperature sensor with chalcogenide glass sensing element. JOAM. 2019;**21**(1-2):48-53

[19] Brodsky MH. Progress in gallium arsenide semiconductors. Scientific American. 1990;**262**(2):68-75. www.jstor.org/stable/24996684. DOI: 10.1038/scientificamerican0290-68

[20] Showa Denko KK. Gallium phosphide (GaP) semiconductors. AZoM; 2019

[21] Starkiewicz J, Allen JW. Injection electroluminescence at p-n junctions in zinc-doped gallium phosphide. Journal of Physics and Chemistry of Solids. 1962;**23**:881-884. DOI: 10.1016/0022-3697(62)90145-2

[22] Handbook of Optoelectronics (Two-Volume Set). Dakin JP, Brown RGW. editors. In: Technology & Engineering. CRC Press; 2006. 1680p. DOI: 10.1201/9781420012019